Cleaning Up the Great Lakes

Great
Lakes

Cleaning Up the

From Cooperation to Confrontation

TERENCE KEHOE

ᴍ

NORTHERN ILLINOIS

UNIVERSITY

PRESS

DEKALB

1997

© 1997 by Northern Illinois University Press

Published by the Northern Illinois University Press,

DeKalb, Illinois 60115

Manufactured in the United States using acid-free paper

All Rights Reserved

Library of Congress Cataloging-in-Publication Data

Kehoe, Terence

Cleaning up the Great Lakes : from cooperation to confrontation / Terence Kehoe.

    p. cm.

Includes bibliographical references and index.

ISBN 0-87580-225-7 (alk. paper)

1. Water—Pollution—Economic aspects—Great Lakes
Region. 2. Environmental policy—Great Lakes Region.
I. Title.

HC107.A14K45 1997

363.739'406—dc21   97-13939

                    CIP

Portions of chapter 5 appeared in "Merchants of
 Pollution?: The Soap and Detergent Industry and the
Fight to Restore Great Lakes Water Quality,
1965–1972," *Environmental History Review* 16, no. 3
(fall 1992): 221–46. Used with permission of the
publisher

For My Parents

Contents

Acknowledgments

I t is a great pleasure to thank all of the individuals and institutions who helped in various ways in the completion of this book. My friend and mentor Austin Kerr has been a fount of valuable advice since my first days in graduate school. John Burnham and Mansel Blackford read early versions of the manuscript and offered useful criticism. A seminar with William Childs was invaluable in helping develop the theoretical framework of the study.

I have enjoyed an excellent working relationship with Mary Lincoln and Susan Bean of Northern Illinois University Press. I am especially grateful to Mary for her initial willingness to give serious consideration to a young scholar's manuscript during a difficult period for university presses. The reviews submitted by Samuel Hays and Theodore Karamanski helped me reduce the length of the study while maintaining its strengths. Pippa Letsky edited the final manuscript with great thoroughness.

I have already thanked the staff at the various archives and government agencies I visited, but special acknowledgment must go to Michael Miller and Harold Webster of the U.S. Environmental Protection Agency for their aid in cutting through red tape and making internal agency records available to me. Finally, my research could not have been completed without generous financial support from the Indiana Historical Society, the Bentley Historical Library, the Lyndon B. Johnson Library, the John F. Kennedy Library, the Ohio State University Graduate School, and the Ohio State University History Department.

List of Abbreviations

A/N	Accession Number
AEC	Atomic Energy Commission
BAT	"best available [control] technology economically achievable"
BOB	Bureau of the Budget
BOD	biochemical oxygen demand
BPI	Businessmen for the Public Interest
BPT	"best practicable control technology currently available"
BTUs	British Thermal Units
CARP	Community Action to Reverse Pollution
CEQ	Council on Environmental Quality
DNR	Department of Natural Resources
DOA	Department of Agriculture
DWSPC	Division of Water Supply and Pollution Control
EPA	Environmental Protection Agency
FDA	Food and Drug Administration
FWPCA	Federal Water Pollution Control Administration
FWQA	Federal Water Quality Administration
GLC	Great Lakes Commission
GPO	U.S. Government Printing Office
HEW	Department of Health, Education and Welfare
IJC	International Joint Commission
MSD	Metropolitan Sanitary District (Chicago)
MWRC	Michigan Water Resources Commission
NPDES	National Pollution Discharge Elimination System
NTA	nitrilotriacetic acid
PAAP	Provisional Algal Assay Procedure
PCBs	polychlorinated biphenyls
PHS	Public Health Service
RG	Record Group
SPCB	Stream Pollution Control Board (Illinois, Indiana)
STPP	sodium tripolyphosphate
UPI	United Press International
WHCF	White House Central Files
WNRC	Washington National Records Center
WPCB	Water Pollution Control Board
WPCF	Water Pollution Control Federation

Cleaning Up the Great Lakes

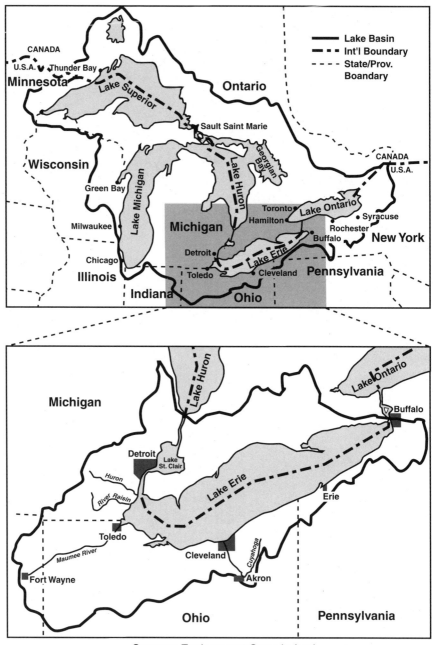

Sources: Environment Canada (top)
U.S. Department of Interior (bottom)

Introduction

The turbulent period of the 1960s and early 1970s generated dramatic changes in American society. Not the least of these changes concerned the expanding role of the federal government in American life and the much greater willingness of many Americans to challenge the authority of established institutions, including the government. During this same period, the modern environmental movement emerged as a powerful force in society, and state and federal governments established the institutions and laws designed to ensure more effective protection of the environment. In this study, I examine the evolution of water pollution control policy in the Great Lakes region and at the national level during these years.[1] My objective is to provide readers with a fuller understanding of the development of the environmental policy system that has held sway in America since the early 1970s. At the same time, I hope to contribute to a greater understanding of broader changes in public life that originated in the 1960s and that have proven to be some of the most important legacies of that era. These include the nationalization of public policy, expanded opportunities for organized groups of all stripes to shape public policy, the breakdown of trust in institutions, and a greater reliance on formal legal mechanisms to resolve conflict and perceived injustices. The emergence of environmentalism helped bring about these changes, and in turn these trends influenced the development of environmental policy-making during the seminal years of the early 1970s.[2]

I have chosen to focus on the Great Lakes Basin for several reasons, aside from the intrinsic importance of this vast system of fresh water. First, studies of environmental regulation have tended to concentrate on the national level; this is understandable, since a major theme of recent regulatory history has been the shift in authority from the state level to the federal government. But prior to 1972, federal water pollution control legislation placed primary responsibility for regulation in

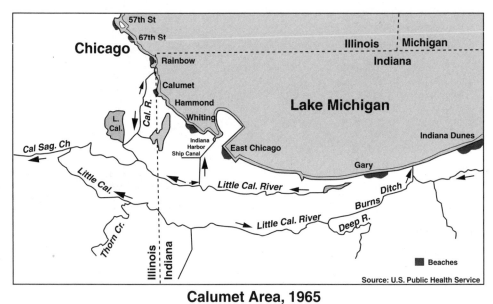

Calumet Area, 1965

the states, with federal officials playing a supporting role. Even after the passage of the 1972 amendments to the Federal Water Pollution Control Act, which placed the ultimate authority for water pollution regulation in the hands of the federal government, state agencies continued to play a vital role in implementing the new legislation and the laws that followed. The situation is similar in other fields of environmental regulation. Thus, state-level policy-making deserves the same degree of attention as federal regulatory efforts. Moreover, the often uneasy relationship between state and federal officials is a critical factor in shaping policy toward the environment.

Second, by concentrating on a particular ecosystem, I can examine the relationship between policy and environmental change in a way that would be unrealistic with a national focus. The Great Lakes Basin is particularly appealing because this region was the industrial heartland of the United States in the decades following World War II. As late as the early 1960s, the states bordering the Great Lakes accounted for just under one-half of the nation's value added by manufacture, and much of their production took place on the shores of the Great Lakes. In 1963, one-third of the top fifteen metropolitan areas ranked by industrial production were located on the Great Lakes. All of the problems associated with the rapid industrial and metropolitan growth of the post-1945 affluent society—especially water pollution—were present to an often extreme degree in the basin.[3] Finally, the problem of Great Lakes water quality makes for a valuable case study because of the sheer size of the lakes and the complexity of the issues involved.

Eight states border the Great Lakes. I can thus make generalizations about state water pollution control policy that would be more tenuous if I were studying the programs of a particular state or an ecosystem that only bordered a few states.

Cooperative Pragmatism

Prior to the 1960s, government regulation of water pollution in the Great Lakes region and other parts of the United States was based on the principles of voluntarism and informal cooperation, administrative expertise, and localism. I refer to this regulatory system, or regime, as "cooperative pragmatism." Cooperative pragmatism evolved out of earlier trends in American government-business relations and environmental regulation.

A major theme in the history of the United States since the Civil War has been the gradual expansion of government involvement in both the performance of the economy and the behavior of individual business firms. Various forms of regulatory activity have been an important component of the expanding role of government in the American economy. In turn, tension between the needs of economic entities (usually business firms) and an abstract concept known as "the public interest" has stood at the center of government regulation of economic activity since the inception of the independent regulatory commission in the late nineteenth century.[4] Until the late twentieth century, most regulatory activity, especially at the national level, concentrated on market structures, pricing patterns, and other issues related to economic efficiency, stability, and equity. In part because of the benefits that could be derived from this type of regulation, representatives of business and other economic interest groups often established close working relationships with particular regulatory agencies. This pattern of cooperation also characterized early efforts to protect public health, such as the regulation of the drug industry.

Despite differences in nuance and explanation, historians and political scientists have developed a rough consensus about the nature of government regulation and the American political economy in the mid-twentieth century. Whether this system is termed "corporatism," "interest group liberalism," "the broker state," or some other appellation, certain common elements predominate. Rooted in the organizational revolution of the late nineteenth century and the Progressive Era, and given impetus by the crises of two world wars and the Great Depression, the new political economy featured an expanded federal government that interacted with organized economic interest groups to shape policy in various areas of concern. Government officials and representatives from organized interest groups—especially business—cooperated through both formal and informal channels, utilizing government agencies, advisory councils, trade associations, and other institutions to sanction and make routine such interaction.[5] In a recent synthesis of work on government regulation, George Hoberg used the phrase "New Deal Regulatory Regime" to describe this system, which also existed in a less complex form at the state level. According to Hoberg, the New Deal regime depended on "norms which placed a great deal of faith in administrative expertise and considered regulatory agencies as legitimate and effective guardians and representatives of the public interest."[6]

Despite Hoberg's emphasis on the New Deal as the creator of this system, a key element—the expertise of the government regulatory official—was rooted in the ideology and reforms of the Progressive Era. Given the central role of free enterprise in the ideology of the American business class and the ingrained reluctance of American business leaders to share their decision-making authority with outsiders, regulatory officials in the United States have always faced a challenging task. In the late nineteenth century and the Progressive Era, reformers and regulatory officials sought to provide these new institutions with greater legitimacy by stressing the expertise of commissioners and administrators and their ability to act as disinterested guardians of the public interest. This ideology of expertise was especially appealing to self-conscious professionals in the emerging fields of public health and conservation, and these ideals continued to influence public officials in these areas long after the Progressive Era had come to a close.[7]

The system of water pollution control that developed in the Great Lakes region in the first half of the twentieth century embodied the faith in expertise associated with the Progressive Era, while also incorporating the cooperative elements described by scholars of the post-1945 political economy. By the 1950s, the authority for regulating water pollution in each of the Great Lakes states rested in part-time boards composed of concerned state department heads and often citizen members chosen to represent affected interest groups, such as industry. These boards established broad policy and made the final decisions on important matters, while the engineering staff from the state health department acted as the administrative arm of the agency. In each of the states, a sanitary engineer acted as the executive secretary and made sure that the board's policies were carried out. Because of the part-time nature of their duties and the technical complexity of much of the work, board members relied a great deal on the expertise and advice of their staff.[8]

The members of the state water pollution control boards and their staff viewed voluntarism and cooperation as the keystones of the water pollution control program and did their best to avoid legal action, especially in the case of industry. As Clarence Klassen, technical secretary of the Illinois Stream Pollution Control Board (SPCB), once put it, "every legal case involving stream pollution indicates the failure on the part of someone, and very often the regulatory agency, to work out a voluntary solution."[9] The state sanitary engineers and most state board members shared a commitment to cleaning up their state's waters, but they also believed in the importance of taking into account other considerations. The engineer-administrators viewed themselves as specially trained, objective experts with the responsibility to balance the sometimes conflicting needs of narrowly focused interest groups. These engineers shunned what they called "the emotional approach" to pollution control and instead tried to chart a pragmatic course that could maintain consensus and make efficient use of economic resources. The engineers and board members in the Great Lakes Basin viewed water pollution as a relative concept and believed in setting waste treatment requirements on a case-by-case basis, taking into account a stream's waste-assimilative capacity, the receiving water's primary uses, economic considerations, and other local factors.

State officials were also determined to maintain regulatory authority at the state level, where local considerations would continue to receive due consideration. Before the 1960s, the federal government's role in water pollution control remained subordinate to that of the states. Officials from Washington sometimes provided technical support, and federal legislation established a modest grant program for the construction of municipal sewage treatment plants. But for the most part the state regulatory officials operated with little interference from their counterparts in the federal program.

The Breakdown of Cooperative Pragmatism

The system of cooperative pragmatism that had developed over decades in the Great Lakes Basin came apart during the 1960s, to be replaced by a regulatory regime that differed in fundamental ways. The changes in water pollution control policy mirrored shifts that were taking place in other areas of regulation and—in a general way—in American society itself. During this period, regulatory institutions and accepted ideas about the best means of controlling various forms of economic activity came under increasing attack from critics who charged that the public interest had been neglected in favor of selfish private interests. Robert Rabin described the years extending from the late 1960s through the mid-1970s as the "Public Interest Era." During these years (self-styled) public interest organizations operating in a number of social policy areas exerted great influence over the formulation and implementation of government policy. New federal laws, often grouped together by scholars as the New Social Regulation, built upon earlier efforts to control corporate social conduct but were unprecedented in their ambitious goals and requirements. The New Social Regulation cut across industry lines and dealt with a broad range of issues, including product safety, corporate hiring practices, workplace safety, and environmental protection.[10]

The nature, as well as the scope, of regulation changed during these years. The federal legislation of the Public Interest Era shifted regulatory authority to the national level and provided federal regulatory officials with greater power to control the behavior of corporations and other relevant entities, including state and local governments. But these laws also attempted to limit agency discretion through firm compliance timetables, precise standards, and other specific guidelines. Thus, while the formal authority of regulatory officials increased greatly, their autonomy or freedom of action actually lessened. Other factors undermining agency autonomy included the greater willingness of the courts to intervene in regulatory decision-making and new administrative procedures that allowed increasingly sophisticated nonbusiness interest groups to influence or block agency decisions. This new system, which Hoberg called the "Pluralist Regulatory Regime," was marked by legal conflict and formal procedures that allowed a greater number of interest groups to influence policy.[11]

A close study of water pollution control in the Great Lakes region reveals the processes at work in this transformation in an important policy area. During the 1960s, increasing levels of pollution, the rise of environmentalism, and entrepreneurial politics at the state and federal level slowly undermined the existing system of water pollution regulation. While the state sanitary engineers and other regulatory officials tried to maintain the system of cooperative pragmatism in the face of these changes, new legislation and regulatory institutions—along with the emergence of a new generation of public officials who embraced environmental values—helped usher in a different era.

In the years following the end of World War II, state regulatory authorities in the Great Lakes Basin succeeded in imposing at least some degree of waste treatment on the effluent discharged by the largest municipalities and industries in the region. In spite of these efforts, the tremendous growth in population and manufacturing activity in the basin placed an increasing strain on the Great Lakes and their tributaries, especially in the highly developed metropolitan belt that extended along the lower shores of Lakes Michigan, Erie, and Ontario. Large oil slicks, beaches closed due to public health concerns, an overabundance of algae—these and other less visible signs of environmental deterioration signaled in the eyes of many the failure of existing regulatory institutions.

The media also began to focus more attention on environmental ills during the early 1960s. The major metropolitan newspapers in the Great Lakes region played a crucial role in supporting the activities of clean water advocates, publicizing pollution problems in the lakes and taking editorial positions that argued for a more aggressive cleanup effort. The expanding media coverage conveyed the message, in both implicit and explicit terms, that not nearly enough was being done to deal with mounting environmental problems. As popular interest in environmental issues began to grow, in part because of the highly visible evidence of environmental decline, the number of people actively involved in organizations concerned with protecting the Great Lakes increased. Samuel Hays and other scholars have portrayed modern environmentalism as part of the broader quest for a better quality of life that grew out of the affluent, consumer-oriented society of post-1945 America. Hays also argued that a strong "sense of place" was often the driving force behind local environmental activism.[12] Sense of place was indeed an important factor in the Great Lakes Basin, where citizen activists and their organizations—usually based in urban centers—sought stronger government action to protect this treasured local resource. Some of the most prominent individuals agitating for stronger regulatory efforts in the early 1960s were average middle-class citizens without a prior history of activism who found themselves driven by their revulsion at the degradation of local water quality to engage in high-profile protest activity.[13]

Politicians and government officials responsible for pollution control—especially at the national level—noted what appeared to be a growing public demand for stronger regulation to protect the environment. In the Great Lakes Basin, clean water advocates and the local media criticized state officials for their willingness to accept degraded water quality in some areas and their unwillingness to engage in formal enforcement actions against major polluters. Whereas state water pollution control officials were by no means passive servants of corporate interests and other pollution sources, the engineers shared common ground with industry representatives in their scorn for the public's "emotional approach" to pollution control and their desire to maintain regulatory authority at the state level. Yet at the same time, the state administrators clearly enjoyed the new prominence given their efforts and found themselves gradually embracing the new attitudes about desired water quality and strict enforcement. But state officials could not change fast enough to suit their critics, who looked to Washington for help and found a receptive ear.

Officials in the Kennedy and Johnson administrations focused on "quality of life issues" such as health care and education as an extension of the traditional liberal agenda. Johnson administration officials eventually embraced environmental protection and enhancement as an important part of the president's domestic program. Younger Democratic members of Congress, including some prominent figures from the Great Lakes states, also seized upon environmental issues as an important area of public concern that deserved greater attention. The combination of bold new legislation and aggressive action by federal agencies continued through the Nixon administration, as the Republican president and his advisers competed with liberal Democrats in Congress for leadership in this increasingly prominent policy area.[14]

In the Great Lakes Basin, throughout the 1960s and 1970s it was the federal government that usually acted as the driving force for a more aggressive pollution abatement effort. During the 1960s, federal officials and local clean water advocates in the region benefited from the activities of each other. Citizen activists and groups such as the Izaak Walton League had attacked the state programs in the past, but they could now turn to a viable alternative for action—the federal government. The resulting clamor for federal action smoothed the way for federal agencies to intervene formally in the region, in some cases at the request of harried state governments. Ironically, the clean water advocates eventually realized that, while federal officials did share their objective of attaining a uniform high level of water quality, federal regulators such as Murray Stein in large part subscribed to the principles of cooperation and consensus in attempting to achieve this objective. Eventually, the system of cooperative pragmatism began to unravel and the public clamor for strong enforcement of antipollution laws steadily increased as earlier efforts failed to bring results.

As a number of scholars have noted, the reformers of the Public Interest Era, including environmentalists, shared a distrust of government bureaucracies. An important component of the ideology that dominated the Public Interest movement—best articulated in the rhetoric of citizen activist Ralph Nader—was the argument that most government agencies had been "captured" by special inter-ests and were incapable of acting in the public interest. This assumption was one reason why legislation enacted during the Public Interest Era sought to limit ad-ministrative discretion and make it easier for citizen organizations to participate in the regulatory process.[15]

Elected officials also responded to public dissatisfaction with government pol-lution control practices by creating new agencies with enhanced capability and less direct participation by regulated interests. At the state level, political ap-pointees took the place of the sanitary engineers in spearheading the new pollu-tion control programs. And just as lawyers assumed the leadership roles in the federal program when it attained new prominence, the new state agencies and departments were often headed by lawyers or other persons without technical training in pollution control. State and federal attorneys general—in particular William Scott of Illinois—also entered the field and sought water quality im-provement through independent legal suits.[16] This new generation of regulatory officials, especially those at the federal level, from the beginning sought to break with the cooperative practices of the past and confront major polluters head-on with formal enforcement action. William Ruckelshaus, the first administrator of the U.S. Environmental Protection Agency (EPA), later explained that his desire to restore public confidence in the commitment and ability of government to take strong action to protect the environment accounted for the emphasis on tough enforcement action during the agency's early years.[17]

The rise of environmentalism shaped the regulation of water pollution in the Great Lakes Basin in an even more fundamental manner. Prior to the 1960s, state control efforts were based on the concept of "reasonable use." According to this doctrine, dischargers possessed a legitimate right to employ a water body for waste disposal purposes, as long as the other primary uses of the waters in ques-tion were protected. In practice, this led to a system of de facto "zoning" in which some rivers and lake areas were allowed to deteriorate and even relatively clean waters were exposed to waste discharges that received less treatment than appeared feasible at the time. Clean water advocates in the Great Lakes region and elsewhere rejected the reasonable use doctrine and instead argued for a pol-icy that required dischargers to apply the maximum level of treatment possible, regardless of the condition of the receiving waters and other local factors. Indus-try officials and state regulators had long scorned this concept as "treatment for treatment's sake," but a growing number of federal officials and members of Congress embraced this ideal. Gradually, national water pollution control policy shifted away from the reasonable use doctrine and toward maximum feasible treatment. The 1972 amendments to the Federal Water Pollution Control Act represented the culmination of this trend, declaring a national goal of completely eliminating all pollutant discharges to the nation's waters by 1985.

Thus, by the early 1970s, the system of cooperative pragmatism that had framed the ground rules for pollution control in the Great Lakes Basin for decades lay shattered, temporarily replaced by a haphazard series of lawsuits and formal enforcement actions directed at polluters by state and especially federal agencies. The authors of the landmark 1972 amendments wanted to continue the more aggressive approach to pollution control, but in a more organized, planned fashion. The new law established a framework of national effluent standards and compliance deadlines that, in conjunction with streamlined procedures for enforcement action, was designed to produce a dramatic improvement in America's waters.

Federalism and Water Pollution Control

The nature of the American federal system changed dramatically beginning in the mid-1960s when Congress began passing laws that preempted the regulatory authority formerly exercised by the states and their political subdivisions. In the area of environmental protection, the sponsors of the new national laws relied on "partial preemption." Under this approach, the state governments were responsible for implementing the national regulatory programs contained in federal statutes. These programs usually involved national minimum standards, which the states were required to adopt. For the first time, the federal government began to regulate directly an important portion of the activities carried out by officials at the state and local level.[18] Even before the enactment of the first federal law to apply partial preemption to water pollution control (the Water Quality Act of 1965), clean water advocates in the Great Lakes Basin were pushing for federal intervention in the region. These activists made common cause with liberal members of Congress such as Gaylord Nelson and Robert Kennedy, who believed that a national system of water pollution control was needed to prevent state competition for industry from undermining environmental protection regulation. Throughout the 1960s and 1970s, Congress and federal officials prodded the states to take more aggressive regulatory action and demand better water quality and higher levels of waste treatment.

State officials strongly resisted the movement toward increased federal authority in pollution control. But whereas the new federal legislation reduced the discretionary authority of the states, the national statutes also provided state governments with powers and responsibility they did not previously possess. These federal mandates also helped spur the movement for the creation of new state agencies with the capacity to carry out increased regulatory responsibilities. The federal laws may have shifted ultimate authority for standard-setting and enforcement to officials in Washington, but the sheer magnitude of the national abatement program ensured the continued reliance on state officials to implement the different aspects of the program.

Large municipalities such as Detroit and Cleveland contributed to the region's pollution problems through the discharge of high volumes of inadequately treated sewage. The regulation of city governments as dischargers by both the federal government and the state governments illustrates another dimension of the federal system. In this case, the regulators' handling of the major municipal pollution sources was complicated by the cities' increasing reliance on federal grants—which were not always forthcoming—to make the required improvements in municipal treatment facilities. For a variety of reasons, the major cities in the Great Lakes Basin were much slower in making abatement progress than their industrial counterparts.

Science, Technology, and Policy

Questions concerning scientific evidence and risk also shaped water pollution control policy. During the 1960s and early 1970s, expanding scientific knowledge about the pollution of the Great Lakes ecosystem forced regulators to deal with new problems and impose new requirements on various groups in the basin. The popularization of ecological concepts such as "the web of life" heightened concerns among some segments of society about the threat posed to humans by the presence of industrial chemicals in the environment. In this way, the focus of water pollution regulation in the United States came full circle. The earliest efforts of government to control water pollution were driven by a desire to protect public health from the threat of waterborne infectious diseases such as typhoid. Innovations in the treatment of drinking water subsequently resolved this problem, and the clean water advocates of the 1960s concentrated primarily on the damage that water pollution caused to the environment. But by the early 1970s, environmentalists who addressed water pollution problems gave more and more of their attention to the toxic pollutants that appeared to threaten human health.

The post-1945 American economy witnessed a tremendous growth in the production of synthetic compounds manufactured with new technologies. Although these advances promised many benefits for society, some scientists and public health officials voiced concern about the toxicity of some of the compounds and their long-term effects on human health and the entire ecosystem. The new synthetic waste compounds differed from those produced by traditional heavy industry in their slower rate of decomposition and their tendency to accumulate in the fatty tissues of fish and animals, thus passing through the food web and ultimately tainting human food. The uncertain health effects of the many new synthetic substances—especially the danger of long-term exposure to low concentrations—drew the attention of influential environmental popularizers such as Barry Commoner. Rachel Carson's attack on pesticide use, *Silent Spring,* when it was published in 1962, became the most famous of the jeremiads against these synthetic compounds.[19]

In the Great Lakes Basin, as in other parts of the United States, efforts to control water pollution from toxic substances revolved around the question of safety thresholds for these compounds. In other words, what level of concentration, if any, could be considered acceptable? This determination was particularly important since it was impossible to eliminate completely the various toxins from the environment. Thus, water pollution control officials found themselves increasingly forced to deal with the murky issue of risk assessment.[20] Because of the relatively recent origin of these compounds and the lack of conclusive data on their effects, fixing threshold values was a tricky business. Regulators found the control side of the equation equally difficult, since many of the toxins entering lake waters did not stem from easily pinpointed production points but from general use of particular products, such as insecticides and PCBs (polychlorinated biphenyls).

Nontoxic pollution problems such as waste heat and eutrophication (nutrient over-enrichment) also forced regulatory officials to confront issues of scientific uncertainty. In these cases, reputable scientists adopted different positions and presented evidence and drew conclusions that directly contradicted their counterparts on the other side of the issue. Regulatory officials were then faced with a choice of either delaying decisions pending further study or choosing a course of action that weighed the potential harm of not imposing new controls against the economic costs and other effects of additional regulation. Ultimately, such decisions came down to a value judgment. The inability of regulatory officials to quickly adopt and maintain a firm policy in these cases further undermined their authority as experts.

≈

Policy making in the United States has become an increasingly complex endeavor in the last decades of the twentieth century, involving more groups, more regulations, more levels of government, and—not surprising—more conflict. The contested field of environmental policy is a striking example of this trend. The growth of environmental regulation has generated vigorous debate about the overall costs and benefits of these laws and the seriousness of the threats posed by various environmental problems. Critics have also focused on the environmental regulatory system itself, questioning the effectiveness and economic efficiency of existing laws and procedures. One goal of this study is to provide for scholars and other interested observers a deeper understanding of the historical development of the contemporary regulatory system. My assumption is that a greater understanding of the evolution of the current system can only improve the quality of the debate.

Part 1

Cooperation and Consensus

CHAPTER ONE

"A Matter of Reasoned Cooperation"

Water Pollution Control through the Early 1960s

In the late nineteenth century, the rapid growth of an urban industrial complex along the American shoreline of the lower Great Lakes generated increasingly serious local pollution problems. Industrial wastes were the first to attract the attention of reformers, but typhoid epidemics and outbreaks of other waterborne diseases shifted attention to the problem of untreated municipal sewage discharges. At first, public health officials took steps to require cities to treat their sewage before discharge. Such regulation was short-lived, however, for the development of effective water supply purification methods in the early decades of the twentieth century undermined support for building costly waste treatment facilities.

In the decades between the two world wars, state regulatory officials began to give greater priority to protecting the environment from the effects of water pollution, but the state programs remained relatively weak and ineffective. This situation changed after 1945. New state laws enacted in the immediate post–World War II period provided the state water pollution control boards and commissions with broad authority and enforcement powers. This legislation recognized the need to place greater restrictions on the discharge of industrial waste and gave new priority to protecting waters for recreational uses. Congress also enacted a major water pollution law. But prior to the 1960s, the federal government played a limited role in water pollution regulation and confined its duties for the most part to technical support and modest grants for the construction of sewage treatment plants.

The sanitary engineers who administered the state water pollution control programs in the 1950s and 1960s took pride in their professional training and expertise. Men such as Clarence Klassen, Blucher Poole, and Loring Oeming worked hard to ensure that significant municipal and industrial dischargers provided at least some level of waste treatment for their effluent. But the engineers' professional outlook, in combination with the economic concerns of their state governments, produced a system of water pollution regulation based on voluntarism and informal cooperation. Most important, the engineers and other state regulatory officials followed the practice of linking individual discharge requirements to a series of local variables, especially the dominant uses of the waste-receiving waters. In each case, it was up to the engineer administrator and his staff to assess the situation, balance the competing needs of various interests, and adopt the appropriate treatment requirements.

Waterborne Disease and Pollution Control

Urban-industrial pollution first emerged as a significant public policy problem in the United States in the second half of the nineteenth century, largely as a result of the rise of the great industrial cities following the end of the Civil War.[1] The rapid population growth of American cities was especially striking in the Midwest, particularly along the shores of the Great Lakes. In the period between 1880 and 1890, the population of Chicago more than doubled, expanding from just over 500,000 inhabitants to more than 1 million. During that same period, Cleveland grew by 63 percent and Detroit and Milwaukee by 77 percent.[2] The concentration of industrial activity created serious nuisance conditions in many cities. Early water pollution studies sponsored by state and municipal authorities often pointed to industrial wastes as the most important cause of declining water quality.[3] In 1886, for example, the Ohio Board of Health reported that during periods of low rainfall the Cuyahoga River, which drains into Lake Erie, received so much waste from Cleveland industries along the banks of the river that "there results a stagnation and deposit, with decomposition and effluvia, causing the boatmen and others engaged in business along the river nausea and often vomiting."[4]

The problem of industrial water pollution faded in significance, however, when the massive program of sewerage construction that began in the 1870s and the ensuing outbreaks of waterborne disease made municipal waste control a much higher priority for public health officials. City officials viewed the waste-carriage system of human waste removal as a great improvement over the traditional methods of individual storage and private scavenger service removal. Unfortunately, the sanitary reformers who pushed for the adoption of this new technology did not foresee that removal of sewage in large volumes to rivers and lakes would lead to frequent outbreaks of typhoid and other waterborne diseases.[5]

Sanitary experts understood the link between bacteria and disease; this was one of the main reasons they had advocated the waste-carriage system. The fatal flaw in their thinking concerned the ability of flowing water to purify itself. According to this theory of dilution, flowing water in a sizable lake or river dispersed and purified sewage through natural processes. Thus, sewage could be discharged into large bodies of water that also served as a source of drinking water without threatening public health. The experts were correct to a point, but they erred in exaggerating a stream's capacity to assimilate unlimited amounts of waste. Students of waste disposal eventually understood that a stream purifies itself through the actions of naturally occurring bacteria that utilize free oxygen to consume organic waste in a process sometimes referred to as *oxidation*. A stream's capacity to purify itself is thus a function of the dissolved oxygen present, and this supply of oxygen can be exhausted until it is replenished. In other words, there was a finite limit to how much organic matter a lake or river could assimilate at one time.[6] Chicago was an excellent example. As the city's population swelled, so did the volume of raw sewage entering Lake Michigan. Chicago soon experienced one of the highest typhoid fever death rates in the country: 159 per 100,000 population in 1891.[7]

During the first decade of the twentieth century, state legislatures enacted a new wave of state water quality laws, as public health officials and other professional groups responded to soaring rates of waterborne disease by lobbying for more stringent state pollution control measures. Severe typhoid epidemics in 1904 and 1905 lent a sense of urgency to the reformers' arguments. Existing laws placed the responsibility for regulating waste discharges on the state boards of health, but lack of funding and inadequate enforcement powers prevented these laws from having much impact.[8] New statutes in a number of states, including Minnesota, Ohio, and Pennsylvania, granted the state boards of health enforcement powers to ensure the protection of public health from contaminated water. Pennsylvania's Purity of Waters Act, passed in 1905, empowered the state health commissioner to order the halt of sewage discharges into state waters, and Ohio's Bense Act of 1908 gave state health officials the authority to require the purification of sewage.[9] The new state legislation appeared to herald a new approach to the control of water pollution.

At the same time these laws were being passed, sanitary experts were developing new technology for the purification of drinking water supplies. The widespread use of filtration and chlorination proved successful in dramatically reducing the incidence of waterborne infectious diseases.[10] One result, however, was that the crisis atmosphere behind the enactment of the stringent state water quality laws was dissipated, and the push for aggressive regulation of waste discharges lost urgency. In effect, public officials made a decision to deal with the problem of water pollution through the treatment of water supplies rather than through the treatment of waste effluent.[11] Sanitary engineers played an important role in this decision, eventually winning over public health physicians who had argued for the need to treat sewage discharges even though water supply treatment practices were gradually eliminating the threat of waterborne disease.

Many sanitary engineers resented the growing power of public health physicians, who dominated the state public health boards. In the early years of the sanitary movement, the boards lacked significant power, and the leading sanitary engineers had been able to advise their municipal clients in sewage procedures with little interference.[12] During the 1890s, public health physicians and sanitary engineers were united in their call for sewage purification as a means to prevent the outbreak of waterborne disease. Now, however, cost-conscious engineers and their municipal clients argued that the effectiveness of water filtration rendered expensive sewage treatment measures unnecessary. It was more cost effective and practical, in their view, for all cities and towns to treat their drinking water, rather than attempting to provide effective treatment for the sewage of thousands of cities and towns. Besides, cities had other health needs but only limited budgets.[13]

Prominent conservationists and other citizens who considered themselves supporters of progressive reform joined the physicians in continuing to advocate sewage treatment. These allies from outside the medical field broadened the argument for waste treatment by emphasizing not just health concerns but the need to protect and preserve the nation's natural resources. Speaking at Buffalo in 1910, ex-president Theodore Roosevelt demanded that the purity of the Great Lakes be protected, declaring that "civilized people should be able to dispose of their sewerage in a better way than by putting it into drinking water."[14] Sanitary engineers responded to these arguments by portraying public health officials as unrealistic or even radical and emphasizing the practicality and financial responsibility of their own position. In the first two decades of the twentieth century, continued improvements to water supply purification methods (such as the addition of chlorination) increased the reliability of drinking water treatment and undermined physicians' warnings about the need for the treatment of sewage. Cities that constructed water treatment facilities experienced steep declines in their typhoid mortality rates. By 1920, the sanitary engineers had won out on sewage treatment, and most public health officials accepted the futility of continuing to press for widespread sewage purification.[15]

The First IJC Study of Great Lakes Water Pollution

The problem of waterborne disease hit the cities located on the shores of the Great Lakes particularly hard. Cities by oceans had to seek their potable water from inland sources, whereas river cities were allowed to withdraw their water from upstream and discharge their sewage downstream. Lakefront cities confronted the problem of polluting their own water source when they discharged their wastes, however. The contaminated state of drinking water drawn from the Great Lakes led to periodic outbreaks of typhoid fever, with the years 1903–1904 and 1909–1910 being particularly severe. In the latter outbreak, typhoid claimed some two thousand victims.[16]

Chicago's solution to the problem of contaminated drinking water was unique among American cities. Chicago's burgeoning population forced the city to take action in the early 1890s, before other urban centers in the region. Because effective water supply treatment had not yet been developed, city officials decided to attack the problem by ceasing all sewage discharges to Lake Michigan. City engineers reversed the flow of the Chicago River and, using water from Lake Michigan, drained Chicago's sewage through the specially constructed Sanitary and Ship Canal into the Illinois River and ultimately into the Mississippi River. Work on the canal began in 1892. The strategy paid off: construction of the sanitary canal and the subsequent diversion of the city's sewage away from Lake Michigan reduced the typhoid death rate from an average of 67 per 100,000 population in the 1890s to 14 per 100,000 by 1910, after the channel system had been completed. Chlorination of Chicago's drinking water further reduced the typhoid death rate to 1 per 100,000 by 1919, the lowest rate in the country.[17]

For a variety of reasons, the other large cities on the Great Lakes were incapable of duplicating Chicago's engineering feat. Continuing outbreaks of typhoid eventually prompted demands for international action to resolve the problem. On March 5, 1912, the U.S. Congress passed a joint resolution calling for the referral of the issue to the International Joint Commission (IJC) for study and recommendations.[18] Three years earlier, U.S. secretary of state Elihu Root and ambassador James Bryce of Great Britain signed the historic Boundary Waters Treaty, which created the IJC for the purpose of resolving issues and disputes arising from the use of the two nations' boundary waters. The IJC was composed of three Canadian and three American commissioners, with one person from each group acting as a cochairman. Common concerns about navigation, lake water levels, water allocation, and similar issues provided the impetus for the treaty, but Article 4 stated that the "boundary waters and waters flowing across the boundary shall not be polluted on either side to the injury of health or property on the other."[19]

The IJC issued its final report on Great Lakes water quality in 1918. The report stated that by far the greatest volume of the lakes' waters existed in a state of "pristine purity." The situation was very different, however, in the shore waters near cities, where the discharge of untreated sewage from both municipalities and shipping vessels caused a significant decline in water quality. Pollution was particularly severe in the waters of the connecting channels between the lakes. Significantly, the IJC barely mentioned industrial waste. The commission did warn that, although industrial effluent was not at present an important contributor to water quality problems, waste from industrial sources could have a very injurious effect in the future if preventive measures were not taken.[20]

The IJC's most important recommendation was that all sewage being discharged to the boundary waters should receive some degree of purification treatment. But in its prescription for determining the level of treatment required, the IJC's position incorporated the "practical realism" of the leading sanitary engineers who had served as consultants during the study. The commission recommended that the level of sewage treatment be determined by the treatment capacity of a water-purification plant. In other words, the degree of contamination in the receiving waters should not exceed the level beyond which a water-purification plant could render the water safe for drinking.[21]

With this position, the IJC sanctioned an approach to water pollution control that focused not on preserving water quality, but on maintaining a level of purity sufficient to ensure safe drinking water after treatment at a water-purification plant. Since water-purification technology was still relatively new (and improving steadily), the practice of linking water quality to the effectiveness of water supply purification methods did not bode well for the future of the Great Lakes. The IJC also recommended that the commission be given sufficient authority to regulate and prohibit waste discharges into the Great Lakes, but concerns about national sovereignty prevented any action on the matter.[22] As the incidence of typhoid fever outbreaks began to recede with the spread of effective drinking water purification, the cities of the Great Lakes, like the rest of the nation, for the most part dealt with water pollution by protecting the drinking water supply.

Water Pollution Control between the Wars

During the 1920s, the views of sanitary engineers continued to shape water pollution control policy. Wisconsin was a pioneer in the establishment of a professional sanitary engineering staff within the state health body. In 1911, Wisconsin officials created the position of state sanitary inspector to oversee the functions of the State Board of Health, and this office was reestablished in 1919 as the Bureau of Sanitary Engineering. Other states soon followed Wisconsin's example and created their own sanitary engineering bureaus or divisions within the state health authority.[23]

The sanitary engineers also shaped policy through the Conference of State Sanitary Engineers, which was composed of the chief sanitary engineers from the state boards of health. Formed in 1920, the conference held annual meetings, and its reports were published in the U.S. Public Health Service's *Public Health Bulletin*. By the mid-1920s, the conference, through its Committee on Sewage Disposal and Stream Pollution, was urging states to enact legislation that would require state approval of new municipal and industrial discharges and provide state authorities with adequate powers to regulate existing waste discharges. Sanitary engineers continued to advocate the full use of a stream's waste-assimilative capacity and a "practical" approach to setting waste treatment requirements. But the proceedings of the Conference of State Sanitary Engineers in the 1920s reveal a movement away from a reliance on water supply purification as the main tool in dealing with water pollution.[24]

In general, sanitary engineers adhered to a set of values that resembled closely the values of conservation professionals in areas such as forest management and river development. Sanitary engineers, like conservation professionals, believed that scientifically trained, objective experts were the most fitted to make decisions concerning resource utilization in the best interests of the nation or community as a whole. These attitudes remained deeply rooted in the profession.[25] The Conference of State Sanitary Engineers was part of an effort by leading sanitary engineers to define themselves as highly trained professionals—"Public Health Engineers"—on an equal level with public health physicians. In the long run this push for broader responsibility and authority was unsuccessful. But sanitary engineers remained powerful forces in the establishing and implementing of public policy in the areas of effluent disposal and water supply treatment.[26]

During the 1920s, state governments began to make tentative movements away from an almost exclusive focus on public health concerns in water pollution control. This reflected the growing interest in outdoor recreation among influential segments of the population. Water resources were being used increasingly for swimming, fishing, boating, and other recreational activities. The sale of outboard motors, for example, increased by 150 percent from the period 1920–1924 to the period 1925–1929. While the number of people actively involved in conservation causes was quite small compared to the later environmental movement, conservation and sportsmen groups such as the Izaac Walton League—founded in Chicago in 1922—made up a vocal lobby in support of tougher pollution laws and more aggressive enforcement of those already on the books.[27] In the early years of the century, some states passed legislation designed to protect fish and wildlife by regulating or prohibiting both municipal and industrial wastes, but various loopholes and lack of enforcement rendered these laws ineffective. The conservation commissions created by some states in the 1920s, including New York and Wisconsin, were intended to cooperate with state health departments in controlling water pollution, but the two agencies often clashed over priorities and the taking of enforcement action against polluters.[28]

To eliminate the confusion and inefficiency caused by the division of regulatory responsibility, some of the states experiencing serious water quality problems centralized water pollution control authority within a single agency. Pennsylvania pioneered in this trend with the establishment of the State Sanitary Water Board in 1923. The board was made up of interested department heads or their representatives. The Department of Health, through its Bureau of Sanitary Engineering, administered the water board's policies. In 1927, Pennsylvania legislators decided to add three private citizens to the board. The use of an interagency board or commission, usually including some citizen members, became the model for state water pollution control in much of the United States and in all of the states bordering the Great Lakes. Illinois, Wisconsin, and Michigan established government bodies similar to the Pennsylvania Sanitary Water Board before 1930, and each of the remaining Great Lakes states followed suit over the next several decades.[29]

In spite of these organizational changes and a renewed emphasis on the negative environmental effects of water pollution, lack of enforcement powers or an unwillingness to use them, insufficient program funding, and a lack of broad public support continued to impede state water pollution control programs throughout the 1930s and 1940s.[30] The policy of stressing water supply purification over effluent treatment, coupled with increasing urbanization and industrialization, resulted in growing miles of contaminated waterways. In 1930, although 87 percent of America's urban population relied on sewers to disperse their waste, only 26 percent had their sewage treated.[31] Industrial wastes also received minimal treatment. State health departments were reluctant to take enforcement action against industrial polluters because of concerns about limiting economic growth, while polluting industries used their political leverage to weaken pollution-control legislation and impede the enforcement of existing statutes. State authorities, rather than relying on formal orders and legal action, stressed the importance of cooperation in dealing with industry.[32]

State public health officials were most concerned about industrial wastes when these pollutants represented a threat to public water supplies. Organic industrial wastes consumed the dissolved oxygen in waterways, thus reducing a stream's capacity to assimilate the infectious bacteria from human waste. For this reason, sanitation experts adopted the practice of describing industrial wastes in terms of their biochemical oxygen demand (BOD) population equivalents. For instance, the waste from a paper mill would be described as having a BOD population equivalent of 100,000. This measurement would signify that the mill's waste used up as much oxygen as the raw sewage discharge from a community of 100,000 people. Industrial wastes could also pose a more direct threat to drinking water supplies. In some cases, industrial pollutants not only caused offensive tastes and odors in tap water but impeded the efficient operations of treatment plants.[33]

Protection of drinking water supplies lay behind interstate conflict in the heavily industrialized Calumet region that straddled the Indiana-Illinois border on the shores of Lake Michigan. In accord with its earlier strategy of effluent diversion, the city of Chicago reversed the flow of the Calumet River, where some major industries were sited, in order to keep these wastes out of Lake Michigan. Unfortunately, the industries across the Indiana border continued to pump large volumes of effluent into the lake, contaminating Chicago's water supply in the process. In the early 1930s, threats of legal action by the state of Illinois forced oil refineries located in Indiana to reduce the amounts of phenols in their wastes.[34]

Ten years later, similar problems forced the state of Illinois to bring suit against the state of Indiana and four of its cities in the U.S. Supreme Court to stop the contamination of Lake Michigan by the municipal and industrial wastes discharged from the Indiana section of the shoreline. The Supreme Court appointed a Special Master to oversee the case, but no testimony was ever taken. Illinois's drastic action forced Indiana dischargers to embark on an extensive abatement program, with the bulk of the money paying for industrial waste treatment improvements. Even after the program was completed in November 1950, at certain times of the year the city of Chicago still had to take additional measures at its South Side filter plant to control taste.[35]

Strangely enough, it was the Great Depression that acted as the spur to a much accelerated rate of sewage treatment plant construction, since the erection of these plants made ideal public works projects. In addition, the attention that the Roosevelt administration paid to water resource planning gave pollution reduction a higher priority than it might have had. The results were impressive: the population served by sewage treatment facilities in the United States grew from more than 21.5 million in 1932 to more than 39 million by 1939. During the same period, the sewered population increased by nearly 8 million, representing a significant total decrease in the discharge of raw sewage. These efforts, however, had little impact on the discharge of industrial wastes.[36]

The Postwar Drive for Clean Waters

In the Great Lakes region, state legislatures enacted significant new water pollution control legislation during the period from the mid-1940s to the early 1950s. The new laws established the general policy that would be followed in each of these states until the mid-1960s. Many other states outside the Great Lakes region also enacted water quality legislation in the immediate postwar period. Several factors were behind this trend. First, the war industry boom and related growth in urban population had a noticeable effect on water quality in many areas. Second, the specter of federal intervention encouraged states to put their own houses in order. The conservation interests and their allies had been lobbying for federal legislation since the 1930s, and Congress finally enacted the first major federal water pollution control law in 1948. Finally, a more subtle factor may have been the general drive for a higher quality of life that followed World War II. Not only was there a greater demand for environmental quality from some segments of the population, but state health department officials, with communicable diseases seemingly under control and community sanitation standards at an all-time high, now felt justified in focusing greater attention on other threats to public health, including various types of environmental pollution.[37]

After the enactment of the state water quality laws in this period, authority for water pollution control in all of the Great Lakes states rested in an interdepartment body that followed the Pennsylvania model. That is, a state board or commission composed of the relevant department heads or their representatives established broad policy and made the final decision on important matters, while the engineering staff from the health department acted as the administrative arm of the agency.[38] Except for New York and Wisconsin, each of the states also included private citizens as members of the board. In some states, the law specified that a private member must represent a particular interest, such as industry or municipal government; in other states, such representation was not required, but state authorities usually viewed it as desirable to have interested groups represented, even if informally. The composition of the Michigan Water Resources Commission (MWRC) was typical. By law, it was made up of the health commissioner, the director of conservation, the highway commissioner, and the director of agriculture, along with three private members representing industry, agriculture, and conservation groups, respectively.[39]

In the states that had already established such bodies prior to the war (Pennsylvania, Illinois, Wisconsin, and Michigan), the new legislation provided these agencies with enhanced authority and enforcement powers. When the other states established their new regulatory authorities, they also endowed them with more extensive powers than the health departments had possessed. The new statutes made it official policy to protect all reasonable uses of public water, including recreation. This policy marked a decisive break with the public health tradition of concentrating almost exclusively on bacterial waste that might threaten water supplies. A new emphasis on controlling industrial waste was a related development, since many industrial pollutants carried no bacteria but did threaten aquatic life and ruined rivers and lakes for recreation.[40]

International and Interstate Cooperation

The city of Detroit initiated the second study of Great Lakes water pollution after World War II when city officials charged that phenolic wastes discharged into the St. Clair River from various oil refineries on the Canadian side were contaminating Detroit's source of water supply at the head of the Detroit River. In April 1946, the U.S. and Canadian governments directed the IJC to make a study of pollution in the waters connecting Lake Huron and Lake Erie (the St. Clair River, Lake St. Clair, and the Detroit River). The two governments later extended the reference to cover the other connecting channels.[41]

With waterborne diseases under control, this investigation lacked the urgency of the original study, but the IJC's final report, issued in 1950, confirmed suspicions about the decline of water quality in some heavily populated areas. IJC investigators found that because of population growth the bacterial concentration in some areas was from three to four times greater than it had been in 1912 at the time of the IJC's first study. Fortunately, advances in water supply treatment methods had virtually eliminated typhoid fever in the region. The IJC found that 96 percent of the population living along the connecting channels was served by sewer systems. Whereas the waste from 86 percent of the population received primary treatment, "only a minor percentage" was subject to secondary treatment.[42]

Primary treatment referred to physical manipulation of effluent, such as screening and settling, that removed less than 50 percent of organic pollutants. Secondary treatment used various methods to re-create and accelerate the oxidation process that occurred naturally in streams and could achieve more than 90 percent removal of organic pollutants. In addition to these standard methods, additional processes—sometimes grouped under the heading "tertiary treatment"—could be incorporated to attain even higher levels of organic pollutant removal or to remove other forms of pollutants, such as phosphorus. The adoption of these methods could often increase treatment costs significantly.

In contrast to the situation in 1912, the IJC reported that industrial wastes were now "a major problem." Industrial wastes discharged to the connecting channel waters made greater oxygen demands than the aggregate domestic waste discharged; and chemicals, suspended solids, and other inorganic industrial pollutants harmed the waters in other ways.[43] The situation was most pronounced in the Lake Huron–Lake Erie connecting channel. The IJC reported that in these waters "wastes containing oils, greases or tars have fouled bathing beaches, coated swimmers, and caused destruction of wildfowl. . . . oil coats the hulls of boats and docks and creates a fire hazard."[44]

To cope with the mounting pollution problems, the IJC listed a number of general and specific objectives for the waters in question and recommended that the two governments adopt the objectives and implement the appropriate remedial measures, as described in the body of the report. The commission recommended specific objectives for particular kinds of waste. But as a general objective, the IJC stated that all types of waste should receive sufficient treatment to protect all legitimate water uses. The governments of the United States and Canada approved the commission's recommendations, but for the most part left implementation to the state and provincial governments.[45]

The IJC appointed advisory boards to monitor progress and advise the commission on all matters related to achieving the abatement objectives for the connecting channels. The IJC and the advisory boards confined their activities to the collection of data and the making of recommendations to the responsible American and Canadian regulatory authorities. Since the advisory boards consisted of the concerned state, provincial, and federal officials from the United States and Canada, the recommendations carried great weight. The state agencies responsible for the connecting channels used the IJC's recommended objectives for the boundary waters as guidelines in establishing requirements for dischargers in these areas.[46]

The environmental strains generated by concentrated populations and industrial production were often most evident in waters that crossed state lines or that were shared by a number of states. Some state governments looked to formal interstate cooperation as a means to resolve issues that arose out of these situations. In 1936, the states of New York, New Jersey, and Connecticut became the first state governments to establish a formal, legally binding interstate compact exclusively for the purpose of controlling interstate pollution. Other states copied this model; in 1948, the states of Illinois, Indiana, Ohio, West Virginia, New York, Pennsylvania, Virginia, and Kentucky came together to form the Ohio River Valley Water Sanitation Compact. Some of the interstate commissions possessed enforcement powers, but in these cases individual states retained a veto power over commission actions. In practice, the interstate commissions performed the role of coordinating member-state efforts and establishing common standards.[47]

In 1955, Illinois, Indiana, Michigan, Minnesota, and Wisconsin entered into the Great Lakes Basin Compact and created the Great Lakes Commission (GLC). By 1963, all the states in the basin had joined the compact. The desire to capitalize fully on the opening of the St. Lawrence Seaway provided the initial impetus for the Great Lakes Basin Compact.[48] Unlike other interstate water basin commissions, pollution control would be only one of the GLC's concerns.

The GLC created no structure for the coordination of pollution control policies other than a standing committee on pollution abatement. As a result, it had little impact on the regulation of water pollution in the Great Lakes.[49]

Federal Support for State Pollution Control

In the post-1945 period, new federal laws gave Washington a much larger role in the control of water pollution, but the states retained primary responsibility for regulating waste discharges within their borders. Prior to 1948, the national government had little involvement in water pollution control, with federal officials confined mainly to various research activities. Federal financing stimulated the boom in sewage treatment plant construction of the 1930s, but the primary goal of the New Deal initiatives was unemployment reduction and economic "pump priming." For critics of state water pollution control, the concrete results stemming from the federal construction program illustrated the potential benefits of an expanded federal role in pollution control and gave new momentum to calls for national legislation.

The outbreak of World War II prevented the passage of such a law, but in 1948 the 80th Congress enacted the Federal Water Pollution Control Act, which placed responsibility for the program in the Public Health Service (PHS). The compromises necessary for passage of the law rendered the act largely ineffective in promoting more rapid pollution abatement. With the economy beginning to boom, Congress was reluctant to spend money on sewage treatment plant construction. And concerns about maintaining state autonomy in the regulation of waste discharges checked proposals for significant federal regulatory authority.[50]

The authors of the Federal Water Pollution Control Act amendments of 1956 sought to provide the federal government with a more significant role in pollution abatement. The most important provisions of the 1956 act were the creation of a federal construction grant program for municipal waste treatment plants and a revision of the procedures for federal enforcement action. The new legislation authorized an annual appropriation of $50 million for construction grants to the states. One-half of the annual appropriation was to be designated for communities with populations under 125,000, while a $250,000 ceiling was placed on grants for any one project. These restrictions limited the program's effectiveness in major urban areas, since the cost of waste treatment plants for large cities could cost many millions of dollars. Still, unlike the failed loan program, Congress consistently appropriated each year most of the money authorized for the grant program.[51]

The provisions for federal enforcement in the 1956 act merit detailed discussion because they established the basic framework for federal water pollution control enforcement action until the passage of the 1972 amendments. The new provisions in the 1956 act required the use of enforcement conferences in resolving interstate pollution problems. The conference approach was designed to maintain state primacy and encourage cooperation between all the parties involved.

Under the 1956 amendments, the surgeon general could call a federal en-forcement conference only at the request of an aggrieved state or when studies pointed to the existence of water pollution originating in one state and endan-gering the health or welfare of persons in another state. Representatives from the PHS and the local, state, or interstate agencies that were involved acted as the conferees. The industries and municipalities responsible for the discharges were given a chance to tell their side of the story at these conferences, and any other interested parties also had the opportunity to testify. The federal enforcement conferences were open, informal affairs, with no testimony offered under oath. After all parties had been given a chance to state their views, the conferees adopted a set of recommendations for the polluters. The recommendations car-ried no force of law, and it was left to the state agencies to make sure that the rec-ommendations were carried out by the offending parties.

After a minimum of six months had elapsed, the surgeon general's immediate superior—the secretary of Health, Education, and Welfare (HEW)—could, if he or she decided no progress had been made in the area, move to the next level of the enforcement process: the convening of a hearing board, which was to be composed of at least five members, including one representative from each state involved and a representative from the Department of Commerce. The law spec-ified that HEW officers or employees could not form more than a minority of the board members. The deliberations of this public hearing were more formal and required testimony under oath. But even if the hearing board decided insuf-ficient progress had been made, the only result was more recommendations. An-other six months then had to elapse before the HEW secretary could begin the fi-nal stage of the enforcement process: initiating proceedings against the recalcitrant polluter or polluters in a federal court. But before the federal govern-ment could take this final step, the HEW secretary had to gain the permission of at least one of the states involved in the action. Penalties against the offending polluter were then left entirely up to the judge.

As the enforcement provisions of the 1956 act indicate, the federal govern-ment's role in the abatement of water pollution remained firmly subordinate to that of the states, although many state officials reacted with alarm to the expand-ing federal role. Like the 1948 act, the 1956 amendments acknowledged the "primary responsibilities and rights of the States" in controlling water pollution. Federal officials during this period continued to reassure state officials that Washington had no desire to usurp state prerogatives in this area.[52]

Cooperative Pragmatism: State Regulators and State Regulation

In the Great Lakes states, the members of the water pollution control boards and their staff viewed voluntarism and cooperation as the keystones of the water pollution control program. The enforcement powers possessed by the individual boards varied from state to state, but in practice all of the boards, because of their shared philosophy, took a similar approach to pollution control. State sanitary engineers worked closely with their industrial counterparts to resolve pollution problems without resorting to formal legal mandates, and the engineers based effluent treatment requirements on the waste-assimilative capacity and primary uses of the receiving stream.

The twin themes of voluntarism and cooperation appeared often in the public statements of state officials during the immediate postwar period when the state governments restructured their water quality programs. When the government of Ohio launched the administration's water pollution abatement program after enactment of the 1951 state law, Ohio attorney general and later governor William O'Neill, writing under the heading "Enforcement versus Cooperation," emphasized his desire to avoid enforcement action, arguing that "much more is accomplished by cooperation than by litigation, and that the time and money spent in the latter is not wise public policy unless all efforts to work together have been exhausted."[53] The state water pollution control boards possessed considerable freedom of action, at least in the strict legal sense. As a contemporary observer noted, one common factor in the postwar state water quality legislation was the "increasing tendency to vest broad discretion in the administrative agency, the legislature stating the policy of the State and leaving it to the agency to determine the most suitable means for accomplishing it."[54]

The state water pollution control authorities in the Great Lakes region differed somewhat in their composition and legal powers, and each operated in a distinct political environment. But the boards shared a number of general characteristics. In each of the states, a sanitary engineer drawn from the state health department acted as the executive secretary and made sure that the board's policies were carried out. The boards usually met once a month, sometimes less often. At these meetings, board members listened to staff reports, issued abatement orders and permits, approved plans for the construction of treatment facilities, met with concerned outside parties, and made various kinds of policy decisions. Because of the part-time nature of their duties and the technical complexity of much of the work, board members relied a great deal on the expertise and advice of their staff. A scholar studying water pollution control in Wisconsin found that the members of the State Committee on Water Pollution rarely interfered with the work of the staff and in fact discouraged staff members from turning too often to the committee for guidance. In Ohio, a visitor from the PHS observed that Water Pollution Control Board (Ohio WPCB) members rarely questioned the recommendations of the staff.[55]

Three sanitary engineers who enjoyed the respect of their peers across the nation were Clarence Klassen of Illinois, Blucher Poole of Indiana, and Loring Oeming of Michigan. Klassen and Poole were the longtime technical secretaries of the Illinois SPCB and the Indiana SPCB, respectively, while Oeming served initially as chief engineer and then executive secretary of the MWRC. The backgrounds and attitudes of these men were typical of their counterparts in the other

Great Lakes states. Klassen and Oeming earned engineering degrees from the University of Michigan in 1925 and 1928, respectively, and Poole took his engineering degree from Purdue University in 1931. Klassen and Poole were generally acknowledged as among the leading water pollution control administrators in the nation. The two received numerous awards and served as advisors and members of many national commissions and committees concerned with public health. Klassen also served as an engineering consultant to the World Health Organization, where he worked with many governments in developing sanitation programs. Oeming was less conspicuous, but still well respected in the profession.[56]

Conscientious men in horn-rimmed glasses and dark suits, the sanitary engineers were professional regulators in the sense that James Q. Wilson used the term. That is, success in their field—in terms of both career advancement and recognition from professional colleagues outside the agency—depended on demonstrating technical competence and behaving in accordance with well-established professional norms. These professional standards were reinforced at gatherings of the Conference of State Sanitary Engineers, the state and national water pollution control associations, and other professional organizations. A strong adherence to these professional norms made it difficult for the state sanitary engineers to adjust their behavior when external pressures mandated changes in regulatory style.[57]

The longevity of the engineer-administrators added to their influence. As civil servants, these men outlasted the appointed citizens and department heads on the boards. Klassen set the record for longevity, serving as technical secretary of the Illinois Sanitary Water Board from 1935 until 1970, when the board was absorbed into the new Illinois Environmental Protection Agency (EPA). Poole served as secretary of the Indiana SPCB from 1945 until his retirement from the State Board of Health in 1970. Their counterparts in the other Great Lakes states also served long terms.

Although the state boards possessed the power to issue legally binding orders, most of the board members preferred to obtain from dischargers informal commitments that they would take the necessary steps to reduce pollution. And when orders were issued, board members usually did all they could to avoid turning to the courts to obtain compliance. This approach was especially pronounced in the case of industries. In 1965, Poole testified that in the Calumet region of northern Indiana, the state's most heavily industrialized and polluted area, the Indiana SPCB had issued only two formal orders to industry—in both cases the relatively small Lever Brothers plant discharging to Wolf Lake. No penalties had ever been collected from industry in the Calumet area.[58] The New York WPCB, except for some orders directed at duck growers on Long Island, issued no orders against New York businesses during the entire decade of the 1950s.[59]

Economic concerns explain much of the reluctance to take more forceful action against industry. Wisconsin had one of the more aggressive programs in the region, but a member of the State Committee on Water Pollution explained, "we on the committee see our work as serving the best interests of the most people.

This means we can't stop pollution completely. It would hurt the state too much economically. We have to try to balance things out and we can't satisfy everybody. . . . Whatever our legal power or technical ability, economics won't let us do such a thing." Government officials in the Great Lakes states worried that tough water pollution regulation might drive industry out of the state. In Wisconsin, paper industry officials sometimes used this threat when faced with strong regulatory action.[60]

William Ruckelshaus, the first administrator of the federal EPA, served as a deputy state attorney general in Indiana in the early 1960s. Ruckelshaus was assigned to the State Board of Health, where he worked closely with both the SPCB and the Air Pollution Control Board. He later recalled that when he and the SPCB pushed hard for compliance with state laws, they would sometimes receive a call from the governor's office, reminding them that the offending industry might relocate outside of Indiana if pressed too much. According to Ruckelshaus, concern about employment usually overshadowed concern about the environment, unless there was some blatant threat to public health involved.[61]

One reason that state regulatory officials were able to rely on cooperation and voluntarism, and be reasonably satisfied with the results, was that their whole approach to pollution control was predicated on making only very reasonable demands of dischargers. While state sanitary engineers and most state board members shared a commitment to cleaning up their state's waters, they also believed in the importance of taking into account other considerations. Clarence Klassen spoke for other sanitary engineers when he argued that lawmakers and administrators had to balance the needs of the interest groups involved in water pollution questions. According to Klassen, these "factions" tended to view the pollution problem "from the angle of their own particular and special interests." In contrast, it was the duty of the responsible administrator to take into account all the relevant factors when making decisions about waste control. Klassen, like other sanitary engineers, favored a flexible approach to pollution abatement that did not attempt to apply uniform water quality standards to all state waters; the unique characteristics of each situation needed to be considered.[62]

Klassen believed that the primary use of the waters in question and the stream's assimilative capacity were the most important considerations. For Klassen, pollution was a relative concept: "Pollution as it affects water quality management is objectionable only in relation to the intended use of the water." Klassen also believed in the full use of a water body's capacity for diluting and purifying waste. This philosophy was known as the "reasonable use" doctrine. To put it simply, a party possessed the right to make reasonable use of a stream for waste disposal as long as the other primary uses of the waters in question were not adversely affected. Thus, the degree of treatment required for a particular discharger was a function of water use, the stream's capacity for waste assimilation, and other factors such as the economic impact of pollution abatement on the community.[63] Any other approach struck the engineers as wasteful and inefficient. Testifying at a congressional hearing on water pollution control legislation in 1963, Loring Oeming challenged the opening policy statement of the proposed law that called for "keeping waters as clean as possible." For Oeming, such a statement was meaningless; the only practical test of pollution control effectiveness was whether other users of the water were protected from interference with their pursuits.[64]

One of the most important areas of policy-making in water pollution control was the classification of water bodies and the establishment of waste treatment requirements. Within this system of classification, there are several terms that need elucidating. (1) Formal *water quality standards* assigned specific numerical levels of quality to particular water bodies or sections of water bodies. In other words, a standard would contain maximum or minimum allowable limits for particular substances, such as coliform bacteria or dissolved oxygen. The standards that regulatory officials applied, except for certain minimum standards applicable to all waters, were usually a function of that water's primary uses. Water classified as a drinking water source, for example, had much higher standards than water classified as primarily a source of agricultural and industrial process water. (2) *Water quality criteria* referred to the standards required for a particular use. Only a minority of Great Lakes states relied on formal water quality standards embodied in state regulations, but every state classified water bodies on an informal basis in order to determine the level of waste treatment required of dischargers in that area. (3) *Effluent standards* were waste treatment requirements that regulators established for particular classes of pollutants. For example, all industries would be required to keep their discharges of phenols below a certain level. Because of their belief in approaching each discharge source on a case-by-case basis, most of the water pollution control officials in the Great Lakes states avoided this kind of standardization, although they all made use of informal guidelines for treating certain pollutants.

Pennsylvania pioneered the use of stream classification and water quality standards in the United States. In the 1920s, the Pennsylvania Sanitary Water Board adopted a policy that divided state waterways into three classes: those unpolluted from artificial sources; those so severely polluted as to be unfit for water supply, fish life, or recreation; and those falling between the first two classes. The board's policy was to protect those waters in the first class from further degradation, while working to improve those waters in the third class. As this policy evolved, it became the practice to set waste treatment requirements for dischargers based on the classification of the receiving stream. Thus, industries discharging to badly polluted waters faced lower treatment requirements than a plant beginning operation in a relatively undeveloped area.[65]

New York was the only other state in the Great Lakes to adopt a policy of establishing formal, codified water quality standards for all of its waters, but the other state authorities followed a policy of taking into account a stream's primary uses when establishing treatment requirements. Aside from the uses to be protected, the degree of dilution afforded by the stream was the other major factor in establishing treatment requirements. In the Detroit area, for example, Michigan state officials generally required municipal treatment plants discharging to tributaries of the Detroit River to provide secondary treatment for their sewage, whereas treatment plants along the Detroit River itself could rely on the river's larger assimilative capacity and thus had only to provide primary treatment. Of course, local economic conditions sometimes became the overriding concern, as when a small town lacked the necessary tax base to pay for a treatment plant or an aging industrial facility served as a town's major employer.[66]

While the state regulatory authorities believed it was their role to balance the competing needs of various interest groups, they also attempted to build a consensus around their pollution control policies by providing private groups with at least a limited role in shaping policy. The allocation of board seats to representatives of private interests and the extensive use of public hearings were part of this strategy. Typically, the boards convened public hearings to provide a forum for discussion of state policy in a particular drainage basin. These hearings provided officials with an opportunity to justify their decisions about water quality standards or treatment requirements and gave affected parties and interested citizens a chance to question and criticize board decisions.

In New York, the WPCB used this approach in setting water quality standards. State engineers first made a broad survey of the drainage basin in question, and the board then released a set of proposed water quality standards. Testimony at the subsequent hearings did sometimes result in the alteration of the proposed standards. In the Lake Erie–Niagara River Basin, for example, the board proposed originally that the heavily industrialized lower section of the Buffalo River be assigned a Class D rating. This was a lower quality rating that required the water to be suitable for agricultural and industrial process water. At the hearing, however, industry representatives argued for the assignment of a Class E rating, which applied to waters suitable only for waste disposal and transportation. They emphasized the "excessive costs" that would be involved in trying to meet the higher standard and the lack of space along the river for installing new treatment facilities. Anselmo Dappert, the board's executive secretary, found these arguments persuasive and recommended that the board members adopt a Class E rating for the lower Buffalo River, which they did.[67] Conservation interests were also sometimes able to influence policy. When local conservation groups raised an uproar over the board's initial classification of the most developed section of the Niagara River, the board compromised with a special rating that took into account some of their objections.[68]

The state programs also utilized advisory committees on both an ad hoc and a semi-permanent basis. In Ohio, the severely degraded condition of the Cuyahoga River prompted Ohio Department of Health officials in 1963 to suggest the formation of an advisory committee to assist the state in its cleanup of the Cuyahoga River. The resulting Cuyahoga River Basin Water Quality Committee was composed of representatives from the major industries and municipalities on the river, as well as health department sanitary engineers. Among other functions, the committee worked with state sanitary engineers on the development of water quality standards for the river. At a May 1965 meeting of the advisory committee, Earl Richards of the health department explained that the Ohio WPCB still subscribed to the idea that the use of a stream for waste disposal was legitimate, as long as it did not interfere with other accepted uses. The problem was to determine what uses should be protected, and state officials looked to the Cuyahoga River Basin Water Quality Committee for help in that task.[69]

Postwar business leaders in the Great Lakes region and throughout America embraced the ideology of cooperation and voluntarism espoused by government regulatory officials. Writing in 1956, a student of water pollution regulation described "the change in industry attitude from one of hostile resistance to one of active cooperation" as one of the key developments in recent years. In the Great Lakes states, regulatory officials echoed these sentiments.[70] Industry officials in the Great Lakes realized that it was in their best interest to develop a close working relationship with regulatory officials and encourage the regulators' reliance on voluntarism. At a conference on the enforcement of water pollution laws in November 1965, John A. Moekle, associate counsel for the Ford Motor Company, commented that in Michigan legal enforcement of water pollution regulations relating to industry had "not been a matter of combat or litigation, but a matter of reasoned cooperation, a working out of problems through mutual efforts of government and industry."[71] And like the state sanitary engineers, industry representatives argued that "reasonable" use of a river's or a lake's waste-assimilative capacity was in the public interest, as opposed to the wasteful doctrine of "treatment for treatment's sake."[72]

Industrial representatives also stressed the necessity of maintaining authority for water pollution control at the state level. Left unsaid was the fact that because of their firms' greater economic importance and political clout at the state level, business executives simply found it easier to exert influence over state officials as opposed to federal regulators. As the threat of federal intervention increased, corporate officials in the Great Lakes states emphasized the close working relationship they enjoyed with state water pollution control authorities and the progress that had been achieved through these mutual efforts.

Lack of funding and staff was also an important factor behind the state programs' reliance on voluntarism and cooperation. After passage of the 1956 Water Pollution Control Act, federal program grants helped stimulate state spending in this area, with total state appropriations for water pollution control increasing from just over $4 million in 1956 to approximately $10 million in 1963. But even with this increase, state expenditures in the Great Lakes region and the rest of the United States remained below what most experts considered adequate.[73] Lack of personnel obviously limited oversight of municipal and industrial treatment operations. The Indiana SPCB justified its reliance on persuasion and voluntarism in part because this approach enabled the board to "maintain better relations with offenders which results in a greater likelihood of effective operation of the pollution abatement facilities with a minimum of supervision by State-employed personnel."[74]

State water pollution control officials valued the technical and financial support that they received from the federal government. State regulators in the Great Lakes enjoyed a good working relationship with PHS technical personnel, and the two groups sometimes made joint calls on local industries. In New York, executive secretary Anselmo Dappert credited the work of a PHS industrial waste expert for much of the WPCB's success in establishing an abatement program in the Lake Erie–Niagara River Basin.[75] But state authorities in the Great Lakes Basin were also determined to retain decision-making authority over water pollution in their waters. In their view, federal intervention was only justified in extreme cases where a state government had clearly been negligent in its duties and interstate cooperation had broken down.[76] Of course, the states welcomed federal financial aid programs, but some administrators in the Great Lakes states voiced concerns about the ultimate impact of such programs, given the relatively small amount of money that was actually made available. In 1958, Poole argued that the federal construction grant program for the building of municipal treatment plants actually delayed action in many cases, because cities or towns that were unsuccessful in obtaining grants often postponed the necessary work in the hope of obtaining aid at a future date.[77]

The Kalamazoo River: A Case Study in State Regulation

When a serious and visible industrial pollution problem developed or became intolerable for various reasons, state authorities engaged in a bargaining process with dischargers. State officials sought a resolution to the problem that would eventually improve water quality but not push the polluters to the point that legal action was required or firms threatened relocation. Close examination of a particular case illustrates how this process worked.

During the 1950s, the organic waste discharged by the thriving Kalamazoo River Valley paper industry placed a heavy burden on the Kalamazoo River, which drains into Lake Michigan. In the early 1950s, the paper mills in or downstream from the city of Kalamazoo installed primary treatment facilities, but the ten- to fifteen-mile stretch of the river below Kalamazoo continued to experience degraded water quality, especially during periods of low rainfall. In May 1958, the MWRC met with Kalamazoo River Valley paper mill officials to discuss the commission's recent survey of Kalamazoo River water quality. The MWRC study found that even with all cities and mills in the area meeting the commission's waste treatment requirements, during periods of low flow the river could only assimilate about 40 percent of the total effluent loading—far less than was acceptable. The MWRC report called for significant reductions in BOD from the mills' discharges to the river, but "an agreement was reached to allow a reasonable period of time to attack the problem by cooperative action." Thus, instead of imposing a deadline for the pollutant reductions or even for submission of abatement plans, the MWRC allowed the paper mills an unspecified amount of time to work out their own solutions to the problem.[78]

The paper mill managements were aware of the significant capital and operating costs associated with the installation of secondary treatment, so they initially pursued other options. A joint industry-city study committee hired a firm of consulting engineers to study the possibility of improving stream quality by constructing upstream reservoirs and releasing water to the river during periods of low flow. The consultants concluded, however, that it would actually be less expensive to provide secondary treatment of paper mill wastes.[79] Two different paper industry study committees drawn from paper mill officials in the river valley also studied possible treatment methods, but again the industries sought to avoid the additional costs of higher levels of treatment. In this case, the two committees also investigated the possibility of restoring water quality by introducing air back into the Kalamazoo through artificial means. The latter experiments produced some positive results, but the MWRC did not wish to pursue the idea.[80]

Finally, at its December 1959 meeting the MWRC instructed the Kalamazoo area paper mills to submit by January 1961 their proposed methods for reducing waste loads to the river. All but one of the firms complied with the deadline, and at the March 1961 meeting, after consultation with paper mill officials, MWRC staff presented abatement timetables to the commission members for their consideration. Representatives from the mills and "other interested parties" also attended the meeting and there ensued a lengthy conference on the mills' proposals.[81]

At its April meeting, after reviewing the conference record, the MWRC adopted an abatement timetable that applied to all ten mills. The timetable included deadlines for preliminary steps such as hiring of engineering firms and approval of plans, but the deadline for final completion of all new treatment facilities was September 1, 1964. By October 1, 1964, each of the mills was to be meeting the required effluent limits. MWRC staff worked out a system of treatment requirements that allowed each firm to dispose of as much waste as the river could handle. Since it had been determined that the Kalamazoo River could assimilate only 40 percent of the current total waste load, the state engineers assigned waste load reductions to each plant, based on individual production outputs, that together accounted for a total reduction in daily BOD load of 61 percent.[82]

A forty-two-month construction deadline was fairly generous for plants of this size, but some of the firms balked at the MWRC timetable. In March 1961, Neil Staebler, a member of the Democratic National Committee, informed Democratic governor John Swainson that Arnold Maremont, chairman of the board of Allied Paper and a strong supporter of the Michigan Democratic party, was very concerned about the pressure being put on Allied Paper to reduce its waste loadings to the Kalamazoo River in such a "rapid" manner. Allied, which employed about fourteen hundred people in the Kalamazoo area, was agreeable to the reductions but did not feel the company had the resources to comply with these goals by the MWRC's deadline. Maremont did not think Allied could implement these changes before 1966, and he implied that a more rapid pace could jeopardize the continued operation of the company.[83] Later that year, Swainson received a letter from the executive vice president of the Kalamazoo County Chamber of Commerce, claiming that the area's paper mills would need at least five years to comply with the MWRC's waste reduction objectives. The chamber of commerce official reminded Swainson that the paper industry was the largest employer in the area and warned that Michigan's tough pollution abatement policy was going to drive the paper industry out of the state and into other states wooing industry with low taxes, free plant sites, and other incentives.[84]

In keeping with its policy of avoiding formal orders whenever possible, the MWRC asked the mills to submit signed stipulations (voluntary promises of compliance) incorporating the state requirements. At its June 1961 meeting, the MWRC approved four stipulations, but the remaining six mills in the area either failed to submit signed stipulations or attached unacceptable qualifications to the proposed stipulations. The commissioners agreed to schedule formal hearings for the remaining mills before issuing orders.[85]

Three of the six holdouts subsequently agreed to stipulations incorporating the MWRC timetables. On September 15, 1961, Dr. Ward Harrison, president of Allied Paper, the largest waste producer of the Kalamazoo River Valley mills, appeared before a special meeting of the MWRC that he had requested and announced that the company was ready to sign a stipulation if the final deadline for completion of facilities was moved to May 1, 1966, if Allied provided its own facilities, or August 1, 1966, if the firm connected to a joint treatment facility that was currently the subject of negotiations between the city of Kalamazoo and some of the paper firms.[86] Chief Engineer Oeming informed the commissioners that Harrison provided no information that had not been considered when preparing the original timetables and recommended only an additional three months for Allied's timetable. Oeming also pointed out that any adjustment in the timetable would have to be offered to the firms that had already agreed to stipulations under the original schedule. However, Oeming's superior, Executive Secretary Adams, argued that considerations other than technical factors had to be weighed by the MWRC in deciding on the proposed revisions.[87]

At their regularly scheduled meeting the following week, the commissioners agreed to move the final deadline for compliance with waste-load reductions to June 1, 1966. They instructed MWRC staff to offer these terms to the two firms that were still holding out and the firms that had agreed to the original timetables. All of the Kalamazoo area firms eventually agreed to the revised abatement deadline, with the last one, Hamilton Paper, finally signing a stipulation to that effect in November 1962.[88]

The situation on the Kalamazoo River differed from that of the urban centers on the Great Lakes shoreline in a number of ways. The most obvious was the reliance of the local economy on a single industry subject to intense competitive pressures. Nonetheless, the MWRC's handling of the problem exemplified state water pollution control in the Great Lakes Basin. If anything, the commission was more aggressive and thorough than most state authorities, imposing concrete deadlines for abatement compliance and specifying wasteload reductions. Still, the MWRC was willing to press the paper mills only so far and agreed to extend an already generous abatement timetable even further when faced with vague talk of economic hardship. Even when confronted with delays and Hamilton Paper's refusal to agree to the revised timetable, the MWRC continued to rely on staff negotiations until stipulations could be agreed upon. The Kalamazoo River case was also representative in that Michigan officials based the effluent reduction requirements on the concept of making maximum possible use of a stream's waste-assimilative capacity.

Part 2

The System Challenged, 1960–1968

CHAPTER TWO

"You Alone Have the Answer"

The Path to Federal Intervention in the Great Lakes Region, 1960–1965

uring the first half of the 1960s, the state officials responsible for water pollution control in the Great Lakes Basin found themselves under growing attack from various quarters for failing to make greater progress in the abatement of water pollution. The state water pollution control programs had made considerable strides after 1945 in securing at least some degree of waste treatment for the effluent emanating from manufacturing plants and municipalities. But the tremendous industrial expansion of the postwar years placed increasing strains on the Great Lakes ecosystem. At the same time, the increased leisure time and disposable income of the affluent society resulted in a growing public demand for clean waters for recreational purposes.

State regulatory officials could provide detailed statistics on the millions of dollars spent on waste treatment facilities since 1945, but the important point was that, by the early 1960s, many people—including some in a position to exert influence on policy-making—believed that the pollution in the Great Lakes and their major tributaries was intolerable and that much stronger regulatory measures were needed. The regulatory system of cooperative pragmatism, with its reliance on voluntarism and informal action, only increased the dissatisfaction of clean water advocates, who hungered for visible, dramatic evidence of abatement progress.

As was increasingly the case in a number of policy areas in post-1945 America, activists, newspaper editors, liberal federal legislators, and other persons and groups pushing for stronger water pollution control looked to the federal government as the vehicle for positive change. The federal water pollution control law, as weak as it was, still contained provisions for federal enforcement action that critics of the state programs could look to as a means for securing more effective regulatory action. The federal water pollution control program grew rapidly in size and scope during the early 1960s, as administration officials sought to respond to these demands and capitalize on the growing interest in environmental protection. Murray Stein, the chief enforcement officer of the federal program, and James Quigley, HEW assistant secretary, became highly visible advocates of more effective pollution control.

Clean water advocates in the Great Lakes states, encouraged by the federal officials' talk of higher standards and strong enforcement, urged their governors to request federal intervention in the most polluted areas of the Great Lakes. Local newspapers and federal legislators representing these areas joined in the calls for the convening of federal enforcement conferences. Expanding scientific knowledge about the conditions of the lakes and the sources of their pollution made the state programs even more vulnerable to criticism for failing to cope with significant changes in lake water quality or to effectively regulate major polluters. State regulators, however, remained adamant in their opposition to federal enforcement conferences in their jurisdiction. Because of this aversion to the holding of enforcement conferences and the reluctance of top HEW officials to challenge state prerogatives, intense grassroots public agitation was necessary to force the holding of federal enforcement conferences.

Private citizens such as John Chascsa and David Blaushild, aided by sympathetic members of Congress and extensive media coverage, kept the issue of Great Lakes water pollution in the news and placed pressure on elected state officials to take some kind of action. The efforts of Blaushild and the others paid off. By the end of 1965, the PHS had intervened and held enforcement conferences covering three of the most industrialized and degraded areas in the Great Lakes: the Detroit River, Lake Erie, and the southern end of Lake Michigan. Local clean water advocates hoped that the conferences would be major turning points in the effort to clean up pollution in the basin.

Water Quality and Expectations

By the beginning of the 1960s, state regulators in the Great Lakes Basin could point to considerable achievements in attaining treatment of waste discharges. Pressure from state authorities spurred the construction of sewage treatment plants. In the period between 1952 and 1957, the state of Ohio almost doubled its sewage treatment plant capacity. In 1959, the MWRC reported that the sewage from well over 90 percent of Michigan's population in the Great Lakes Basin received treatment before discharge. The other state boards reported similar progress in their sections of the basin.[1]

State water pollution control authorities in the Great Lakes Basin also succeeded in securing at least some degree of waste treatment for the great majority of industries in the region. The days were gone when an industry could simply, with impunity, discharge untreated wastes into the waters of the Great Lakes. In Indiana, for example, the SPCB reported that 35 of the 37 plants in the Calumet River Basin that discharged to Lake Michigan or a tributary were providing adequate treatment or control facilities for their wastes. In Ohio, 132 out of 198 industries discharging waste to Lake Erie or its tributaries were classified as having adequate treatment, with 53 possessing partial treatment not yet considered satisfactory.[2]

The statistics used by water pollution control authorities in the Great Lakes states to demonstrate their progress appeared impressive on the face of it, but the figures were misleading. In these status reports the state pollution control boards, with the exception of the Pennsylvania Sanitary Water Board, did not usually specify the degree of treatment provided at the facilities (designations such as "adequate" were, of course, quite subjective). This is a crucial factor, since different levels of treatment removed different percentages of pollutants. Much of the municipal effluent discharged to the Great Lakes Basin was subject to only primary treatment, which was far less effective than secondary treatment in removing organic pollutants. The industrial waste treatment processes in use by Great Lakes industries also left significant amounts of pollutants in the final effluent stream.

The incomplete nature of waste removal in the basin was of vital importance, since the region experienced tremendous growth in the postwar years. The manufacturing centers of the Great Lakes Basin boomed in the decades following World War II, as their factories, mills, and refineries struggled to keep pace with the demands of the emerging affluent society. Between 1947 and 1963, the value added by manufacture more than doubled in each of the major metropolitan areas on the shores of the Great Lakes and more than tripled in Rochester.[3] The population in the American portion of the Great Lakes Basin also expanded rapidly during this period, with most of the increase occurring in the suburbs of the largest metropolitan areas. Although the major cities of the region had entered into a period of population decline or stagnation by the early 1960s, their suburbs continued to grow in size.[4] As a result of this manufacturing and population growth, even though a greater percentage of sewage and industrial waste was receiving some kind of treatment, the total waste load to the Great Lakes Basin was steadily increasing.

At the beginning of the 1960s, the status of Great Lakes water quality was unclear. There had been no comprehensive scientific study of water pollution in the region since the IJC investigation of more than a decade before. The gross nuisance conditions in the industrialized tributaries and along some shorelines were obvious to the casual observer, but the condition of the interior waters was not as easy to determine. The steady decline of commercially valuable fish populations in some of the most productive fisheries in the Great Lakes was one indicator of significant changes in general water quality. Fishery biologists, however, attributed this more to overfishing than to pollution.[5]

State water pollution control officials admitted that significant problems remained, but they were optimistic that the steady progress of recent years would continue and ultimately resolve these problems. Federal officials, on the other hand (perhaps because they had no record to defend or because they desired an expanded role in pollution control), were more willing to acknowledge that abatement efforts had resulted in only limited gains, both in the Great Lakes and throughout the country. In 1959, the deputy chief of the PHS Division of Water Supply and Pollution Control (DWSPC) warned that there was "evidence of increasing and more serious pollution [in the Great Lakes]. Unless vigilance is maintained and remedial measures stepped up, pollution here will grow in volume just as it has elsewhere."[6]

Many citizens in the Great Lakes Basin also did not share the optimism of state officials. Although the most severe water pollution in the Great Lakes was localized to some extent, these degraded areas contained the majority of the American population in the region. The closing of public bathing beaches in Cleveland, Milwaukee, and other cities was a manifestation of water pollution that affected even the most casual recreational user.[7] Despite efforts by state authorities to build consensus, the waste loads associated with rapid urban-industrial growth in the Great Lakes Basin were in direct conflict with the growing demand for clean water expressed by vocal organized interest groups and—increasingly—their allies in government and the media.

The Expanding Federal Program

During the 1950s, American political leaders turned more to "quality of life" issues such as health care and education in their bids for popular support. This trend was particularly strong at the federal level and in the Democratic party. In the 1960s, when the Kennedy and Johnson administrations sought to broaden the liberal agenda by expanding Washington's activities into new areas, one of the fields that administration officials looked to was environmental protection. Neither administration needed much prodding when grassroots and congressional demands for an increased federal role in water pollution control grew progressively louder. As a result, the federal water pollution control program expanded rapidly in size during the first half of the 1960s, growing from a $53.8 million budget and 406 employees to a $125.1 million budget and 1,594 employees between 1960 and 1965. The federal program also increased its enforcement activity during this period, although top officials remained hesitant about intervening in the nation's most industrialized regions.[8]

Even before the Democratic party returned to power in 1961, top PHS officials were emphasizing the degraded condition of the nation's waters and calling for a general rise in the level of wastewater treatment throughout the United States. In his opening address to the National Conference on Water Pollution in 1960, surgeon general Leroy E. Burney called the condition of America's rivers and lakes "a national disgrace." A month earlier, Gordon McCallum, DWSPC chief, warned that "despite our increased efforts, the problem of water pollution is growing worse rather than better." And in a 1959 article, assistant surgeon general Mark Hollis argued that in view of the growing waste load to streams, the growing demand for reuse of waters, and the introduction of new kinds of pollutants, "the concept of assimilation of pollution by streams simply breaks down as a major usable factor in the treatment of wastes."[9]

Legislators in Congress played an important role in the expanding activities of the federal program, both through the drafting of new laws and through oversight activities. A number of liberal Democrats in the House and the Senate took an active interest in water pollution control and pushed for a greater federal role in controlling the problem. Senator Edmund Muskie of Maine and Representative John Blatnik of Minnesota took the lead in sponsoring new legislation; Representatives John Dingell of Michigan and Henry Reuss of Wisconsin and Senator Gaylord Nelson of Wisconsin also called for more federal involvement. As the decade progressed and water pollution became an important issue of public concern in the Great Lakes region, Democrats such as Senator Robert Kennedy and Representative Richard McCarthy of New York and Representative Charles Vanik of Ohio emerged as vocal critics of state water pollution control programs and worked for federal intervention in the region.

Liberal members of Congress maintained that state economic rivalry mandated a truly national system of water pollution control. Senator Nelson complained that when he had been governor of Wisconsin, the state's powerful paper industry had successfully used the argument that stricter pollution control standards would put Wisconsin manufacturers at a competitive disadvantage with firms in more lenient states.[10] Senator Nelson, Senator Kennedy, and other federal lawmakers who championed cleaner water pushed for greater federal involvement and national standards as a means of standardizing pollution control across the states and removing this powerful argument against abatement.[11]

Representative Dingell was perhaps the harshest critic of the federal water pollution control program in the early 1960s. Dingell represented a district that included the southwest section of the city of Detroit and its southern suburbs running along the west bank of the Detroit River. The sometimes caustic Dingell was a fixture at congressional hearings on water pollution, where he regularly scored the PHS for its inaction and ineffectiveness. Dingell charged that the PHS leadership was too focused on narrow public health concerns and did not pay enough attention to the devastating impact water pollution had on recreational opportunities. Dingell argued that an aggressive federal water pollution control program was essential because most state programs were underfunded and understaffed and unwilling to take on major polluters. During the first half of the 1960s, this kind of criticism from Dingell and other legislators was an important factor in moving federal officials to intervene in some areas of heavy water pollution.[12]

After 1961, federal lawmakers interested in pollution control found a sympathetic ear at the White House. In one of his first special messages to Congress in February 1961, President John F. Kennedy addressed conservation issues, giving special attention to the need to protect the nation's water supply for all uses through a more effective national pollution control program. Kennedy's attention to environmental concerns on this and other occasions marked the first time a president had sponsored any direct initiatives on environmental policy since the New Deal. The Kennedy administration also displayed a willingness to exercise federal authority in environmental policy. Unlike his immediate predecessor, Dwight Eisenhower, Kennedy was not hesitant about expanding federal spending in these areas. Eisenhower had vetoed new water pollution control legislation the year before because of the proposed increases in the federal construction grant program, but in July 1961 the new president signed almost the same bill into law.[13]

The Water Pollution Control Act of 1961 significantly increased federal funding for the construction of municipal waste treatment plants, raising the annual authorization from $50 million to $100 million after 1963. The law also contained several important new provisions. Federal enforcement authority was extended to all navigable water bodies in the United States, including coastal waters, which meant that all of the American sections of the Great Lakes were now subject to this authority. The federal government could initiate action against intrastate pollution, however, only at the request of the state governor. The final important provision of the act shifted responsibility for the administration of the federal water pollution control law from the surgeon general to the HEW secretary.[14]

The DWSPC still remained buried within PHS bureaucracy, however, a small unit of the Bureau of State Services. Despite the division's low bureaucratic profile, its administrators were determined to promote greater public awareness of the need for effective water pollution control. The PHS "Clean Water" campaign, begun in 1962 and continued for several years thereafter, received the public service support of the national television and broadcasting industries and the endorsement and assistance of the Advertising Council and the National Association of Broadcasters. The division's Information Branch distributed spot announcements to radio stations and filmed announcements to television stations. At the end of these announcements, interested citizens were asked to write to a Washington, D.C., address to find out what they could do to help combat water pollution. In one five-month period in 1964, these spots generated over sixteen thousand letters.[15]

While the DWSPC Information Branch sold the American public on the importance of water pollution control, HEW assistant secretary James M. Quigley attempted to sell an expanded federal pollution control program to professionals in the field. A lawyer and former congressman from Pennsylvania, Quigley was appointed to HEW in 1961 for service to the Democratic party after losing his seat in Congress. In November 1961, HEW secretary Abraham Ribicoff designated Quigley as his principal representative for water resource programs, including pollution control. In this role, Quigley was responsible for reviewing situations that might require federal enforcement and recommending action to the secretary.[16] Quigley had no experience in water resources issues, but he assumed a high profile in promoting the growing federal water pollution control program. In various public pronouncements, Quigley emphasized the importance of vigorous enforcement in effective water pollution control and the need to protect water quality for as many uses as possible. In an interview shortly after his designation as the HEW secretary's representative on water matters, Quigley emphasized the economic importance of water-related recreation and tourism and promised a more vigorous federal program.[17]

Quigley also delivered his tough talk directly to state officials. In a May 1962 address before a joint meeting of the State and Interstate Water Pollution Control Administrators and the Conference of State Sanitary Engineers, Quigley told the audience that "the best way to keep us out of your state is to get the pollution out of your water. If you can't and want us to help, we will be happy to try. If you don't we may be forced to try even though you don't want us to." In private, members of Congress and their aides made it clear to state officials that if the states did not produce better results, Congress would see to it that Washington took over responsibility for standard-setting and enforcement. Quigley later recalled that "the Public Health Service had become too understanding of the problems of the state administrators, and that the state administrators had in turn been brainwashed to the point where they were too understanding and sympathetic to the problems of the polluters."[18]

The enforcement conferences remained the primary vehicle for federal enforcement action, but they were called infrequently. In the period between the passage of the Water Pollution Control Act in 1956 and the enactment of the 1961 law, the surgeon general called only fourteen federal enforcement conferences, and six were mandatory responses to state requests. For the most part, the PHS avoided heavily developed areas such as the Ohio and Hudson Rivers and concentrated on rural waterways like the Big Blue River (Nebraska–Kansas) and Animas River (Colorado–New Mexico). Despite Quigley's tough talk, the federal government continued to be cautious about convening enforcement conferences even after the HEW secretary assumed direct responsibility for administration of the water pollution control program. Five enforcement conferences were held during all of 1962 and 1963, with three initiated at state request.[19] Critics of the federal program such as Representative Dingell found this inaction infuriating, given the obviously polluted conditions of many interstate waterways. One reason for the hesitancy of HEW officials was the outright opposition to such federal intervention among state officials, except in cases where they requested the convening of a conference. Since the PHS depended on the cooperation of state health agencies for the successful implementation of many important programs, PHS officials tried to avoid antagonizing state authorities.[20]

Once an enforcement conference was held, federal authorities were very reluctant to move to the next stages of the enforcement process: the formal hearing board and then referral to the U.S. attorney general. Of the thirty-seven enforcement conferences held between 1957 and the end of 1965, only four moved to the hearing board stage, and only one of those four resulted in a court case.[21] The rare use of the hearing board and federal suits in part reflected the law's inbuilt delays and safeguards against unnecessary use of federal power. But the reluctance to move beyond the conference stage was also a reflection of federal officials' enforcement philosophy. The men from Washington talked frequently about the need for cleaner water and faster abatement progress. In practice, however, federal regulators pursued an enforcement policy that mirrored to a great extent the cooperative pragmatism characteristic of the state programs.

Murray Stein, chief of the DWSPC Enforcement Branch, served as the chief enforcement officer of the federal water pollution control program from 1956 until the creation of the federal EPA in 1970. As chief enforcement officer, Stein embodied the federal government's style of water pollution regulation for well over a decade. Since it was Stein who chaired most of the federal enforcement conferences, he became the most visible official in the federal water pollution program. The Brooklyn-born Stein spent his entire life in government service. At the age of nineteen, after several years of study at New York's City College, Stein left for Washington, where he worked as a printer's devil in the Government Printing Office before serving as a medical sergeant during World War II. After the war, Stein earned his way through law school at George Washington University by clerking in the law office of the Federal Security Administration, where he was hired after graduating with honors. When the 1956 Water Pollution Control Act created the need for an enforcement officer in the federal program, HEW officials offered Stein the job.[22]

In spite of his background, Murray Stein's personality and appearance did not fit the stereotype of the conservative career bureaucrat. Stein's thick Brooklyn accent, his use of humor, and his penchant for straight talk made him a colorful figure, and he worked to cultivate good relations with the local press in the different cities to which he traveled. Stein's job kept him on the road from two to three weeks a month, but he claimed to enjoy the travel. The chunky Stein, who possessed a fondness for long, dark cigars, obviously relished being the center of attention at the conferences. Stein lacked technical training in water pollution control, but his years of practical experience helped make up for the absence of technical expertise, and he established a solid reputation in the field.[23] During Stein's service as chief enforcement officer, federal authorities continuously pressured the state agencies to accelerate their abatement programs, raise pollution control requirements, and take a more aggressive stance against recalcitrant polluters. Ironically, during most of these years, the federal enforcement program pursued a policy that eschewed litigation and that (like the states') depended on voluntarism and informal cooperation.

Stein was a strong advocate of the "conference approach," which he traced back to a 1921 Supreme Court ruling in a dispute between New York and New Jersey over interstate pollution. In the ruling, the Supreme Court suggested that cooperative study and conference between the states' representatives made a more effective approach than proceedings in court.[24] Stein often pointed out with pride that his agency had been involved in cases concerning thousands of industries and cities but had gone to court only once. Stein said that he and the other federal officials measured their success by the solutions hammered out at the conference table, rather than by the number of court actions they brought against polluters.[25]

The use of an open conference also reflected federal officials' continued emphasis on consensus in pollution control, an important aspect of cooperative pragmatism. Stein believed that the conference, because it encouraged participation by all interested parties, helped develop a kind of community spirit in regard to the problem. According to Stein, the key was to assemble all the relevant facts about a particular pollution situation: "I think when we put these before the public, all reasonable men can agree on what the situation is. Once we do get this agreement, I think the solution presents itself and the community moves forward."[26]

The first major federal initiative on Great Lakes water pollution grew out of a controversy over the effects Chicago's diversion of Lake Michigan water had on water levels and flow rates in other parts of the Great Lakes Basin. In 1960, after the United States intervened in the case, the Department of Justice asked the PHS to conduct a study of water quality in the Lake Michigan–Illinois River Basin to determine current conditions and the effects of returning Chicago's treated effluent to the lake. This study became the initial phase of the Great Lakes–Illinois River Basin Project, a broader comprehensive study of the entire Great Lakes system.[27] During the early 1960s, federal officials placed a new emphasis on the importance of comprehensive planning in water resource management. In the PHS, the DWSPC leadership concluded that such planning could most effectively be carried out by concentrating on individual river basins. The general objectives of the Great Lakes Project and similar programs across the country were to determine the causes and effects of pollution, develop agreement on desired uses and the quality needed to protect those uses, and formulate abatement measures and a time schedule for their implementation.[28]

The work of the Great Lakes Project was carried out in such a way as to gain the participation of all interested parties and (it was hoped) forge a consensus so that there would be general acceptance of the final water quality management plan. Thus, like the enforcement conferences, the river basin planning projects sought consensus among interest groups. This approach was also made necessary by the limited enforcement powers of the PHS. Since the top HEW officers continued to be cautious about the convening of enforcement conferences, PHS regulators had to rely on the voluntary acceptance of pollution control measures by state and local officials and (depending on the aggressiveness of those government authorities) the waste dischargers themselves. Involving these parties in the actual development of the abatement measures would improve the chances of the program actually being carried out.[29]

PHS officials relied on the agency's own resources to carry out the pollution assessment component of the program, but the Great Lakes Project administrators established technical committees to shape the standards and abatement components of the plan. The project staff first set up a technical committee for the Lake Michigan–Illinois River Basin and later established similar committees for Lake Erie and other areas. The Lake Michigan Technical Committee was composed initially of representatives from the four interested state programs, municipal authorities, and area corporations. Later, the technical committee organized subcommittees to propose water quality criteria and standards and included representatives from conservation groups, universities, and research organizations on these bodies. The technical committees later established for other sections of the Great Lakes Basin included a broader representation of interest groups from the start.[30] The state officials participating in the technical committee to the Lake Michigan–Illinois River Basin section of the Great Lakes Project frequently expressed concerns that the federal government was going to use the project to impose its will on the states, although the federal officials directing the project were quick to deny the charge. The state representatives on the technical committee were particularly concerned about the possibility that the federal government might mandate water quality or effluent treatment standards that the states would have to follow.[31]

It was the federal enforcement conference, however, that state officials viewed as the most blatant encroachment into state affairs. Although the recommendations that emerged from the conferences carried no legal force, the possibility of eventual federal legal action against state industries or municipalities was a threat that state authorities could do without. Even more important, state officials viewed the highly publicized conferences as embarrassing ordeals that held state failures up to public scrutiny. And although it eventually became clear that Stein and the other federal officials had no intention of moving beyond the conference level, the convening of a conference gave the PHS the right to hold follow-up sessions to the original conference and examine progress on the original recommendations. Again, most state officials did not relish the thought of a federal Big Brother constantly looking over their shoulders in public as they attempted to deal with complex pollution problems. State officials were also well aware, however, that public dissatisfaction with state pollution control programs created a positive climate for intervention by federal authorities. In 1960, Blucher Poole presented a paper at the annual industrial waste conference at Purdue University. In this paper he warned that "the next decade will probably determine whether the states are going to continue as the dominant force in the control of water pollution abatement or whether they will abdicate to the federal government and the metropolitan areas."[32]

Grassroots Pressure

At the same time that state officials grappled with a growing federal presence in water pollution control, they faced an increasing demand for clean water at the local level. During the first half of the 1960s, organized groups of citizens in the major urban centers along the Great Lakes subjected the state pollution control authorities to a growing barrage of criticism. Traditional conservation groups such as the Izaak Walton League were now joined by good-government organizations such as the League of Women Voters and new single-issue groups that focused all their attention on water pollution. These organizations called for more stringent treatment requirements and more aggressive enforcement action against dischargers. State authorities had always been subject to criticism in the past from local clean water advocates, but these citizen groups now found strong allies in Congress and federal agencies to support their calls for enhanced water quality. Moreover, the major metropolitan newspapers in the region strengthened the political influence of citizen groups by publicizing their activities and raising general awareness of the problem. The largest dailies published articles that emphasized the degraded quality of local waters and editorials that called for an accelerated abatement effort.

There is almost no polling data available prior to the mid-1960s with which to gauge Americans' attitudes toward the environment. But throughout the 1950s, members of the state water pollution control boards in the Great Lakes region and their staff engineers commented on the increasing public awareness of the need to control water pollution. Some of them linked the growing public intolerance for degraded water quality to a desire for the recreational opportunities provided by clean streams and lakes. Indeed, representatives from the Izaac Walton League and other traditional conservation groups used the growing importance of the recreation industry as an argument for more effective pollution abatement. At the National Conference on Water Pollution in 1960, a representative from the AFL-CIO explained that the shortened work week, longer paid vacations, and higher standard of living in postwar America created recreational opportunities for American workers that had been closed to previous generations.[33] Such comments may have been self-serving, but these contemporary observations support the work of Samuel P. Hays and other historians linking the growth of environmentalism to the rise of the postwar affluent society.[34] Although the manifestations of a new environmental consciousness did not become fully visible until the 1960s, there is much evidence from earlier years to demonstrate the growing use of the outdoors as an amenity. For example, overnight visits to state parks increased from 3 million to over 20 million from 1946 to 1960.[35]

Private citizens who became involved in Great Lakes water pollution issues during the 1960s were usually motivated by a strong sense of place. In other words, they were driven by a desire to restore or protect areas with which they were familiar because of residential proximity or frequent recreational use.[36] At some point in their letters or testimony, interested citizens often recounted horror stories of personal encounters with polluted beaches or stinking rivers and recalled how clean they had once been. They expressed concern about the future experiences of their children. Economic self-interest was also a factor. Many who complained about the impact of water pollution were concerned about the negative effect on lakefront and riverfront property values. Fear of waterborne disease was another common element. Some citizens were old enough to have memories of the typhoid outbreaks of the early twentieth century, and the publicity given to the dangers of nuclear fallout and pesticide use made people more aware of the potential health threats posed by substances in the environment. During his testimony before the Lake Erie federal enforcement conference in August 1965,

Cleveland city councilman John Pilch posed the following questions:

> How can you tell if by some youngster swimming in that polluted water, that polluted lake, and he swallows one or two mouthfuls, that maybe years later, he will suffer from some sickness, possibly leukemia? Who can tell? Nobody knows what it is. Possibly cancer—who can tell? Can you? Can I? No, but it could happen.[37]

These citizen groups and interested individuals criticized the principles of cooperative pragmatism that formed the operating core of the state regulatory programs. Clean water advocates rejected the idea of factoring in a stream's assimilative capacity and other variables when determining wastewater treatment requirements and instead called for a policy of maximum treatment. The state officials' emphasis on cooperation also came under fire. Critics pointed out that major problem areas remained obviously polluted and showed little signs of improvement in spite of the much ballyhooed achievements of the state programs. Although state authorities took enforcement action against automatic laundries, dairies, and other small fry, they charged, the major polluters went untouched.[38]

The conservation groups and other state critics hungered for lawsuits and fines, and other evidence that progress was being made. More and more, they looked to the federal government for action against large polluters. In September 1963, Robert Hutchings, chief of the DWSPC Information Branch, reported that "so far as we have been able to learn, the public continues to identify water pollution control progress with the federal Government and the Public Health Service. There is sometimes bitter discontent, however, with local water problems."[39]

The local news media—especially the newspapers—in the urban centers of the lower Great Lakes fueled this discontent with heavy coverage of local pollution problems and editorial stances that usually supported the views of the citizen groups. During the 1960s, the American press became more willing to question and criticize government policy. In the 1950s, many journalists had found themselves unwilling allies of Senator Joseph McCarthy and his cohorts when they followed the convention of merely reporting the official record with no commentary on its accuracy. Government's willingness to cover up and distort the truth in cases such as the U-2 spy flights over the Soviet Union also contributed to growing skepticism among elements of the press, as well as other segments of American society. During the 1960s, younger reporters especially tried to overcome government officials' attempts to manage the news by engaging in more investigative work and treating government policy and official statements with greater skepticism.[40]

In the major cities of the Great Lakes region, a crusade for cleaner water may have appealed to editors and publishers because the pollution issue appeared relatively straightforward and could be presented in nonpartisan terms. In 1960, a survey of newspaper editors from around the country found that, while they were aware of the progress that had been made in controlling water pollution, the editors believed that state programs were not achieving results at a fast enough rate.

The editors called for stronger laws and stricter enforcement of existing statutes, along with higher levels of state funding for pollution control.[41] The reporters who covered water pollution issues, moreover, often took a strong personal interest in pollution control. Robert Drake of the *Cleveland Plain Dealer,* a frequent writer on water resource problems, viewed his role in part as that of a "gadfly of the civic conscience." In a 1962 address to the Ohio Water Pollution Control Association, Drake argued that "gross pollution" was "an offense against human decency" and that there were sufficient laws and treatment techniques to get the job done, if the will and public support were there.[42] Russell G. Lynch, conservation columnist for the *Milwaukee Journal,* later became the first chair of the Wisconsin Natural Resources Board when it was created in 1967.

The newspapers in the lower Great Lakes Basin adopted the sometimes apocalyptic rhetoric of the clean water advocates and foretold environmental disaster unless dramatic action was taken. In a 1962 editorial under the heading "Poisoning the Waters," the *Detroit News* described the decline of commercial fishing in the region and the problems of excessive algae growth and warned that "the Great Lakes could become another Dead Sea."[43] The newspapers eagerly seized upon government reports that examined local or regional pollution problems and highlighted the worst findings. In April 1960, the *Cleveland Plain Dealer* ran two front-page articles on consecutive days that detailed the findings of a soon-to-be-published Ohio Department of Health report on the Cuyahoga River. In the lead article, "Industry Waste Turns River into a Menace," Robert Drake described in grim detail the effects on the river of the more than three hundred tons of industrial waste discharged daily by industries concentrated on the lower Cuyahoga.[44] At one time or another, most of the newspapers in the major cities along the Great Lakes ran special series that examined national, regional, and local pollution problems, while articles in Sunday supplements offered color photos of the worst abuses.

State pollution control officials, who had often lamented the public apathy toward water pollution, at first welcomed the expanded media coverage of their field. When a *Plain Dealer* editorial in March 1960 criticized the Ohio WPCB for not moving fast enough to clean up the state's waters, Ralph Dwork, state health director and WPCB chairman, shrugged off the criticism and commented that "we need such attention on the pessimistic side to help us in a final push against water pollution. We certainly don't want a complacent public attitude."[45] The relentlessly negative tone of the news coverage began to alarm some officials, however. By February 1963, Poole was warning his counterparts in other states that they needed to do a better job of counteracting negative news coverage by informing the public about the progress that had been made, or else people would wonder what their state water pollution control agencies had been doing all these years. Later that year, Loring Oeming presented a paper at the annual meeting of the national Water Pollution Control Federation (WPCF), in which he warned that "the public seems to be demanding a quality of water higher than that which is now accepted by pollution control authorities as providing adequate protection. . . . Aesthetic considerations [Oeming said, and the attitude that] the very best obtainable in sanitary quality is none too good, can be expected to result in gradual elevation of present quality objectives."[46]

Poole was correct about the trend in public opinion. Many citizens living near the most polluted areas of the Great Lakes were expressing doubts about the performance of the state authorities. Pronouncements by federal officials about the need for more thorough waste treatment and tough enforcement action only added to the dissatisfaction of clean water advocates, who increasingly looked to Washington for the answers to regional and local pollution problems. During the first half of the 1960s, critics of the state programs in the Great Lakes Basin sought to involve the federal government more directly in the regulation of waste dischargers by having the PHS hold enforcement conferences in some of the most polluted areas in the region.

The Detroit River

In the metropolitan Detroit area, the closing of a popular state beach dramatized the region's pollution problems. The resulting public outcry eventually resulted in the first federal enforcement conference to address water pollution in the Great Lakes. The Detroit River had long been subject to the strains of servicing the waste load of a massive urban-industrial complex. As the decade of the 1960s opened, a series of events demonstrated in dramatic fashion the polluted state of the river. In March 1960, industrial wastes and sewage caused the deaths of thousands of ducks in the lower Detroit River. This was the first major duck-kill since 1948; investigators found oil on most of the carcasses and autopsies revealed severe inflammation of the ducks' digestive tracts.[47]

An even more important event took place in August 1961 when Michigan authorities were forced to close Sterling State Park Beach on Lake Erie because of contaminated water quality. Ironically, the state of Michigan had either spent or appropriated well over $1 million for the development of Sterling State Park at the time of the beach closing. The Sterling State Park Beach was the only public beach on Lake Erie in the state of Michigan, and the proximity of the park to I-75, the interstate highway connecting Toledo and Detroit, made it easily accessible for citizens living in the region. Attendance at the park had grown steadily in the previous ten years, from just over 100,000 in 1952 to a peak of more than 1,200,000 in 1959.[48]

For residents of the area, many of whom had moved to the lakeshore area because of its recreational opportunities only to confront an increasingly foul lake, the closing of Sterling Beach was the last straw. The negative impact of despoiled beaches on property values was a major concern of interested groups and individuals from the towns and cities of the Michigan section of the Lake Erie coastline. Resort areas along the lakeshore reported a significant drop in visitors because of lake pollution. Many citizens also pointed to the threat of waterborne disease, blaming a recent outbreak of hepatitis in the area on the polluted waters of Lake Erie.[49]

The Lake Erie Cleanup Committee helped focus the outrage over the beach closing and the general degradation of lakeshore water quality. The committee, formed after the Sterling Beach closing and headed by John Chascsa, a councilman from the local community of Estral Beach, concentrated its efforts on lobbying government officials at all levels. But from the beginning, Chascsa and his colleagues looked to the federal government to rescue them. At an early meeting, Chascsa announced a petition drive to get Michigan senator Philip Hart to obtain federal help for the region. Chascsa explained that the cleanup committee was turning to Hart for aid because of the senator's previous work to protect Michigan's recreational resources through the creation of federal parks.[50] Local, state, and federal officials attended the meetings of the Lake Erie Cleanup Committee and briefed citizens on the abatement programs that were already in progress, but by December the committee was completely disillusioned with the lack of positive action from state and local officials. In a letter to HEW Secretary Ribicoff, Chascsa described the committee's frustration with red tape and inaction and declared that "you alone have the answer."[51]

Pressure for federal intervention also came from officials representing cities south of Detroit who complained about contamination of their drinking water supplies as a result of inadequately treated sewage discharged by the Motor City. Officials from the city of Wyandotte were particularly vocal about the threat to their drinking water. The city council passed a resolution in May 1960 asking the PHS to increase its activities in the area. In November 1961, Wyandotte officials sent a letter on the Detroit River situation to Secretary Ribicoff, claiming that median coliform levels in this section of the Detroit River were exceeding the objectives proposed by the IJC in 1950 by 1,000 percent in some months.[52]

Democratic governor John Swainson realized that he had to make some kind of response to the public uproar emanating from southeastern Michigan, but he received conflicting advice from two groups. Prominent Michigan Democrats in Congress urged Swainson to request the convening of a federal enforcement conference on the Detroit River and the downstream Michigan section of Lake Erie, whereas the state officials responsible for Michigan's water pollution control program firmly opposed such action. On October 18, Senator Hart informed Swainson that he had received "countless petitions, letters, clippings, etc., dealing with the water pollution situation" and advised the governor that water pollution control officials in the HEW had informed him they had the funds and personnel necessary to conduct a thorough study of the area and formulate recommendations for corrective action.[53] Milton Adams, MWRC secretary, argued against such a request by Swainson, since it would result in a surrender of state authority in dealing with the complex pollution problems of the Detroit area. Adams was also galled that Michigan would be the first state to request federal intervention to deal with an intrastate problem, and he raised the specter of independent federal legal action against area dischargers. As an alternative Adams proposed an extensive state study of the problem.[54]

Swainson met with Adams, Murray Stein and Gordon McCallum of the PHS, and an aide to Senator Hart on November 7, 1961, to discuss the issue, but the governor could not reach a decision. The following week, one of Swainson's staff informed him that, since the meeting, "our mail has become considerably heavier and all in favor of federal intervention."[55] On December 5, Swainson wrote Secretary Ribicoff requesting assistance under the federal water pollution control law. Ribicoff responded by scheduling an enforcement conference for March 1962.[56]

Michigan clean water advocates hoped that, with the federal government now involved, strong action would be taken to clean the waters of southeastern Michigan. Murray Stein, appearing before the board of directors of the Michigan United Conservation Clubs in Flint, promised forceful federal action. Stein told the conservation leaders that the PHS was not coming in just "to make a report to file on the shelf. We are coming in for one purpose, and that is to clean up pollution and make the Detroit waters available for the maximum number of uses. When we come into a case, we come in to win."[57]

Despite such buildup, the enforcement conference was rather anticlimactic, its major recommendation being a call for further study of the situation, albeit one to be directed by the PHS. At the beginning of the conference, the PHS technical staff reported on their preliminary investigation, undertaken in the months between the announcement of the conference and its commencement. The PHS study was preliminary in nature because of limited data, but the staff did find significant pollution. The levels of bacterial contamination in many areas were well above IJC objectives and posed serious health hazards. The industrial waste data was fragmentary. Significant reductions of industrial waste constituents had been achieved in the Detroit River since the IJC studies of the late 1940s, but despite these reductions, oil, phenols, and other industrial wastes continued to degrade area water quality.[58]

Stein adopted a more cautious stance once the conference was in session. After listening to the PHS report, Stein pointed out that in many parts of the country the pollution problem resulted from industries and municipalities simply discharging their wastes untreated into local waterways. In these cases, the solution was relatively simple: provide some degree of treatment and dramatic improvement would result. Here in Detroit, just about all plants and municipalities had invested in some degree of treatment. Yet there were still significant pollution problems. This and the heavy concentration of population and industry meant that careful study was needed to determine which dischargers needed additional treatment and which were providing an adequate amount. The conferees agreed that the PHS would undertake an extensive study of the area's problems in close cooperation with the concerned Michigan agencies. The conference eventually reconvened in June 1965, when the study was completed.[59]

The one issue of controversy at the conference concerned the degree of sewage treatment that should be provided by Detroit. Since the vast majority of the municipal wastes discharged to the Detroit River emanated from the city's treatment plant, and Detroit provided only primary treatment, implementation of secondary treatment seemed a logical step. But Gerald Remus, general manager of the Detroit water and sewer system, strongly opposed this step. Remus argued that the most immediate need was to connect suburban communities with completely inadequate treatment to the Detroit system. Remus was concerned that if the federal government imposed higher treatment standards on Detroit this would disrupt plans for bringing more communities into the Detroit system because these municipalities might be unwilling to pay for higher treatment.[60]

Whereas Michigan water pollution control officials cooperated with HEW in the holding of the conference and in the subsequent study, they remained bitter about the federal encroachment into their territory. Adams continued to view Swainson's request as a mistake, and the chief engineer of the Michigan Department of Health complained to his counterpart in Pennsylvania about how PHS officials had issued negative public statements prejudging the situation prior to the conference and then had given Michigan no credit for previous pollution control measures.[61] Indeed, it is fair to say that the pride of Michigan officials was hurt by the decision to request federal help. In October 1963, Senator Hart issued a press release crediting the intervention of the PHS for recent water quality improvements mentioned by the federal agency at a briefing in May. Loring Oeming, now MWRC executive secretary, noted correctly that the federal director of the Detroit River–Lake Erie study had actually given all the credit for the improvements to actions previously initiated by state and local authorities. Oeming communicated the MWRC's dissatisfaction to Governor George Romney, who relayed the message to Hart.[62]

If state officials balked at the idea of federal enforcement action, local citizen groups, liberal members of Congress like Senator Hart, and in most cases the local media welcomed federal intervention. The local and national media's increasing attention to the water pollution issue also contributed to a favorable climate for federal enforcement activity.[63] Even more important, the many congressional hearings on the issue that began in 1963 and continued onward created a general pressure for greater federal action. The hearings subjected the federal program to close scrutiny, and the major bills under consideration proposed the creation of a new Federal Water Pollution Control Administration (FWPCA) within HEW but outside the PHS. Some critics of the federal program such as congressman John Dingell went even further and argued that when the new agency was established it should be removed from HEW and transferred to the Department of Interior.

These various pressures resulted in a flurry of activity by HEW secretary Anthony Celebrezze in late 1963. In a two-month period, Celebrezze initiated enforcement proceedings on eight rivers. By contrast, Celebrezze had convened only three enforcement conferences during his first year in office after replacing Secretary Ribicoff in 1962. The greater willingness of HEW officials to employ the federal enforcement power was also a function of the Kennedy administration's realization that a vigorous pollution control effort could pay political dividends. Administration strategists believed that a strong push for clean water could have broad appeal in the 1964 presidential campaign. Thus, President Kennedy himself announced one of HEW's enforcement actions during a "conservation tour" of the states.[64]

The rivers targeted by Celebrezze included the Monongahela and other major waterways, but also some rural rivers in sparsely populated areas. The Menominee River, which drains into Green Bay and forms part of the border between Wisconsin and Michigan, fell into the latter category. Celebrezze's letter announcing the enforcement conference took state officials in both Michigan and Wisconsin by surprise. Neither state was aware of pollution from the other side of the river that was "endangering the health or welfare of persons" on the opposite side, and each state was satisfied with the other's control efforts.[65]

The Menominee was subject to heavy pollution loads from four paper mills, but each state was satisfied with the progress being made by the two mills on its side of the border. In fact, the previous year the Wisconsin State Committee on Water Pollution publicly praised the Kimberly-Clark mill at Niagara and the Scott Paper operation at Marinette for their progress in reducing the pollution of the Menominee River.[66] At the conclusion of the Menominee River enforcement conference on November 8, 1963, James McDermott, assistant attorney general of Wisconsin, presented a statement on behalf of the state conferees arguing that the conference had demonstrated no interstate pollution in the area and that the pollution that did exist was being adequately dealt with by Wisconsin and Michigan. McDermott concluded by stating that "abatement of such pollution cannot be had overnight, but must of necessity involve the passage of sufficient, reasonable time" to allow dischargers to take the appropriate abatement action.[67]

Lake Erie

At about the same time that state officials were dismissing the need for an accelerated abatement effort in the Menominee River, new information about the changing water quality of Lake Erie prompted calls for stronger regulation of dischargers and federal intervention in the region. Disturbing revelations about the biological status of Lake Erie in the early 1960s set the stage for public protest. Given the steady decline of Lake Erie's commercially valuable fish populations over past decades, it was not surprising that the new findings originated in the U.S. Bureau of Commercial Fisheries. Earlier studies of Lake Erie indicated that, in spite of pollution in many of the lake's tributaries and shore waters, the vast interior of the lake remained relatively unaffected. Bureau investigators were thus startled to find that approximately 1,400 square miles of the central basin's bottom waters were almost completely devoid of dissolved oxygen during parts of the year, while other areas suffered from abnormally low oxygen levels. W. F. Carbine, regional director of the U.S. Bureau of Commercial Fisheries, and his colleagues attributed the decline in oxygen levels to the rapid nutrient enrichment of the lake and the subsequent acceleration of the eutrophication process.[68]

Eutrophication is the process whereby lakes become progressively more enriched by a growing supply of plant nutrients. This excessive enrichment results in the rapid growth of algae and other aquatic plants, which in turn results in oxygen depletion as the algae die off and (like any organic waste in water) generate biochemical oxygen demand. The massive algae blooms accompanying eutrophication also contribute to a general degradation of water quality, thus rendering a lake unsuitable for many purposes. Eutrophication is a natural process, but the changes it brings about normally occur over a period of thousands of years. The major sources of human nutrient inputs are municipal sewage, industrial effluent, and farm fertilizer runoff. Phosphorus, carbon, and nitrogen are among the most important nutrients that stimulate algae growth.[69]

When Carbine used the phrase "dying lake" in discussing the accelerated eutrophication of Lake Erie, it fell on fertile ground in northern Ohio as first commercial and sport fishermen, then broader conservation groups, and finally the general press picked up on the phrase as a rallying call to decry the polluted state of the lake. Ohio state officials responded by explaining that Carbine had really meant the lake was experiencing natural changes, albeit at an abnormally rapid rate. The Ohio WPCB issued a statement defending the water quality of Lake Erie, and board chairman Ralph Dwork noted that tests at municipal intakes along Lake Erie indicated conditions had actually improved at a number of points. Dwork did admit that the Cleveland area continued to be a problem, primarily because of the many bypasses in the sewer system that allowed untreated sewage to flow directly into the Cuyahoga River and Lake Erie.[70] The Ohio WPCB's explanations failed to satisfy many concerned citizens, however. Robert Drake of the *Cleveland Plain Dealer* noted, "When a lake is rendered unfit for the legitimate purposes people want to use it for, if it's not dying, it might as well be." The massive deposits of decaying *Cladaphora* (green algae) that washed up on the beaches of Lake Erie that summer added weight to eutrophication concerns.[71]

As Dwork had conceded, the Cleveland area continued to experience severe water pollution, and it was from Cleveland that the loudest cries for abatement action and federal intervention emanated. During the early 1960s, Cleveland's two daily newspapers—the *Press* and the *Plain Dealer*—featured many articles on water pollution and attempted to raise public awareness of the problem. When the Ohio WPCB publicly scolded Cleveland at a meeting in February 1962 for failing to meet all the conditions of its annual discharge permit before issuing the city a new permit after promises of future progress, an editorial in the *Plain Dealer* scored the city for its inaction and expressed the hope that the WPCB "will really crack down a year from now if there is not the promised compliance."[72]

The campaign to clean up the waters of Lake Erie was spearheaded by an unlikely crusader. David Blaushild, a middle-aged Chevrolet dealer in the affluent Cleveland suburb of Shaker Heights, had never been active as a conservationist, but he had once served as a city councilman. Blaushild was an avid fisherman and boater, and he witnessed firsthand the declining quality of Cleveland-area waters. Blaushild was also alarmed by the findings of the Bureau of Commercial Fisheries regarding oxygen depletion in Lake Erie. In the spring of 1964, Blaushild became so disgusted with the situation that he decided to put his marketing experience to work and begin a major publicity campaign to force government authorities to take some kind of action.[73]

Blaushild began his campaign with large newspaper advertisements in Cleveland newspapers. The first ad ran under the heading "Lake Erie is dying: Does anybody care?" and included a coupon that could be returned to Blaushild Chevrolet with the respondent's name and address. Blaushild then sent a petition form to each inquirer with the request that the individual obtain signatures from neighbors and friends. The car dealer planned to forward these signatures to state and federal officials. A later newspaper ad warned of the threat to public health posed by a polluted Lake Erie and quoted from a 1961 federal report that described as uncertain the lake's future status as a fresh water resource. Blaushild's campaign attracted even more attention when he took out an advertisement on a massive 80-by-20-foot billboard overlooking the Inner Belt Freeway near downtown Cleveland. In bold letters, the sign said "Let's *Stop* Killing Lake Erie. Have your council vote Anti-Pollution!"[74]

The billboard's message referred to the model antipollution resolution that Blaushild had authored and distributed to lakefront communities throughout northern Ohio. Between July 6 and August 4, 1964, fifteen communities on or near Ohio's Lake Erie shoreline, most of them in the Greater Cleveland area, passed resolutions modeled on Blaushild's original and then forwarded them to Secretary Celebrezze and other government officials. The municipalities ranged in size from small towns such as Conneaut to Euclid, a large suburb just to the east of Cleveland. The resolutions were rather general, first noting the importance of Lake Erie and its declining water quality, and then calling for immediate enforcement of present antipollution laws by local, state, and federal authorities.[75] Despite the seeming innocuous nature of these resolutions, an attempt to enact a similar resolution in Cleveland generated much controversy.

Mayor Ralph Locher of Cleveland opposed the Blaushild resolution, citing concerns about the impact on local industry of strict enforcement of antipollution laws. Locher was also concerned about the potential costs for the city itself. As was the case in most older cities, Cleveland's "combined sewer" system served as a conduit for both sanitary sewage and water runoff from city streets. When rainfall exceeded a fairly low level, the system became overloaded, and to prevent flooding, both the sewage and rainwater had to be routed directly to the receiving waters, avoiding the treatment plant entirely. This direct runoff of raw sewage from a major city had a devastating impact on water quality, but the massive costs of digging up the original combined system and separating the sanitary and rainwater sewers prevented remedial action. Some experts estimated that the cost of separating Cleveland's sewers would be over $700 million.[76]

On September 3, 1964, representatives from the PHS, the state of Ohio, the city of Cleveland, and local industries met in Cleveland to discuss the current pollution situation and plans for future action. Robert Ferguson of Republic Steel also represented the Cuyahoga River Basin Water Quality Committee. The participants viewed the Blaushild resolution as, in effect, calling for a federal enforcement conference, which they did not believe was necessary. They feared that passage of the resolution might tip the balance in favor of an enforcement conference and so opposed its passage. Those in attendance wanted to continue working through the Great Lakes–Illinois River Basin Project, since it provided them with a voice in planning abatement measures through the technical advisory committee that had been established for Lake Erie.[77] PHS representative Charles Northington, like other regulatory officials, was ambivalent about the activities of Blaushild and other antipollution activists. He scored their emotional approach and what he viewed as distortion of facts. But Northington also believed that the public pressure they provided would act as a powerful impetus for action in this area and make industries and municipalities more willing to cooperate with state and federal regulatory officials.

Discussion of the situation received a public airing later in the month when the Cleveland City Council held a well-publicized meeting on the proposed resolution and the city's water pollution problem. Ferguson of Republic Steel presented a joint report from the Cuyahoga River Basin Water Quality Committee and the Ohio Department of Health that emphasized the $20 million spent on water pollution abatement in the past thirteen years by industries on the river, but Northington observed that the councilmen "were not overly impressed with the big money story given by industry and the administration." The report placed most of the blame for local problems on Cleveland's combined sewer system. Mayor Locher surprised observers by backing the antipollution resolution. He explained that he had only wanted an opportunity to make sure the potential costs were understood and to give industry a chance to tell its side of the story. The city council committee passed the Blaushild resolution and it was later enacted by the full council.[78] According to Northington, the general feeling among the industry and government officials after the meeting was that a federal enforcement action was inevitable. At this point, senior federal officials were trying to remain neutral about the controversy in Cleveland and not get drawn into local politics. Complicating the situation was the fact that Celebrezze had served as mayor of Cleveland for a number of years before becoming HEW secretary.[79]

At the same time that Blaushild and northern Ohio city councils promoted grassroots pressure for federal intervention, members of Congress pushed for action in Washington. In early March 1965, Representative Charles Vanik of Cleveland asked Secretary Celebrezze to convene a federal enforcement conference on Lake Erie water pollution. In the following weeks, Senator Robert Kennedy, Representatives Dingell and Richard McCarthy, and six other members of the House—including some Republicans—submitted similar letters seconding Vanik's request. Assistant Secretary Quigley responded to each request by writing that Secretary Celebrezze could act unilaterally only in cases of interstate pollution. While the pollution of Lake Erie was extensive, Quigley explained, interstate pollution had not been established conclusively under the meaning of the federal law. The department's hands were thus tied, unless a governor from one of the Lake Erie states requested a conference.[80]

By the spring of 1965, intensive media coverage at both the local and the national level had made Lake Erie a national symbol of the nation's pollution problems. In his

Special Message to the Congress on Conservation and Restoration of Natural Beauty in February 1965, President Johnson used the eutrophication of Lake Erie to illustrate the grave environmental problems facing America.[81] The following month, public release of the preliminary findings of the Great Lakes Project study of Lake Erie added to the sense of urgency felt by those urging greater action in the basin. Lake Erie field station chief Northington emphasized the eutrophication problem, significant changes in the biological character of the lake, and high coliform levels in the shore areas near metropolitan areas like Cleveland.[82]

The Ohio WPCB continued to oppose federal interference in the Lake Erie Basin. But Republican governor James Rhodes was feeling the heat and sent several inquiries to Celebrezze about federal responsibility in the basin. Quigley reiterated HEW's stance that the secretary could not convene a conference because of the absence of proven interstate pollution. Quigley informed Rhodes that he had discussed the possibility of an enforcement conference with Celebrezze in response to a governor's request and that the secretary "indicated he would welcome such a request."[83] In other words, the onus was on Rhodes to make the decision.

To demonstrate his concern about the issue, Rhodes held a special governors conference on Great Lakes pollution on May 10, 1965, which was attended by Governor George Romney of Michigan, Governor Nelson Rockefeller of New York, and representatives from other Great Lakes states. Rhodes's opening statement included a call for greater state and federal cooperation, but he did not deal directly with the issue of federal enforcement intervention. Like Governor Rockefeller, Rhodes stressed the urgent need for a much larger federal construction grant program. Other speakers addressed the enforcement conference issue more directly. Representatives Vanik and McCarthy and Senator Kennedy (who submitted a written statement) each renewed his call for the convening of a conference.[84] Members of the Ohio WPCB took an equally firm stance against a conference. Gordon Peltier, Ohio Department of Commerce director and vice chairman of the board, emphasized Ohio's tradition of state responsibility. He argued that a request for federal enforcement would imply either that Ohio could not handle its own problems or that Ohio was attacking neighboring states through HEW and possibly the federal court system. Peltier was also concerned about driving industry from Ohio through overzealous pollution control efforts.[85]

Dr. Emmet Arnold, Ohio director of health and chairman of the WPCB, argued that the water pollution problems of Lake Erie had been blown out of proportion by extremists. There were some polluted areas along the shoreline, Arnold conceded, but the vast interior of the lake possessed excellent quality and was hardly the "giant cesspool" described by critics. He admitted that "mysterious changes" were taking place in the lake such as the tremendous algae growth and shifts in fish populations, but he called for more research into these phenomena before drastic measures were taken. Arnold felt that Ohio had an effective abatement program in place and pointed to the millions of dollars spent by both municipalities and industry to upgrade waste treatment in the basin since the board was created in 1951.[86] At the conclusion of his statement, Arnold, an active sport fisherman, noted that he sometimes enjoyed fishing in the wilderness streams of Canada, but that "it has always struck me that they would be a very poor place to go if you were looking for a job. There is a lot more employment along Ohio's busy waters and we need more and better jobs."[87]

As a savvy politician, Governor Rhodes realized that he needed to avoid the intransigent stance of his department heads. After the conference, Rhodes announced that there was a strong possibility he would ask Secretary Celebrezze to convene an enforcement conference on Lake Erie pollution in Cleveland. On June 11, Rhodes sent a formal request to Celebrezze for such a conference and pledged the complete cooperation of all Ohio officials.[88]

Southern Lake Michigan

In the southern section of Lake Michigan, the massive industrial complex straddling the Illinois-Indiana state line on the shores of the lake made it a relatively easy matter to demonstrate the existence of interstate pollution. Moreover, the Great Lakes–Illinois River Basin Project concentrated its initial study efforts on this area, so that by the mid-1960s there was a wealth of information about the extent of pollution in southern Lake Michigan and the major waste sources. The Chicago water supply system, of course, had been dealing with the problem of Indiana waste for decades. In spite of continuing problems, however, Illinois and Chicago officials appeared content to let the Indiana authorities work at their own pace on abatement in the region. At a 1963 congressional hearing held in Chicago on water pollution in Lake Michigan and the Calumet River system, which contributed much of the industrial waste to the lake, both mayor Richard Daley of Chicago and Clarence Klassen of the Illinois Sanitary Water Board expressed their confidence in the work of the Indiana SPCB and stressed the harmonious working relationship that the two states enjoyed.[89]

The PHS report presented at the hearing undercut these optimistic assessments. The federal study found that the water in the main body of Lake Michigan was "of very high quality," but the southwest section of the lake, especially along the shoreline, exhibited high coliform readings, excessive algae growth, and high phenol and ammonia levels. Since Chicago drained the city's effluent into the Illinois River Basin, the majority of the waste entering Lake Michigan emanated primarily from industrial sources in Indiana.[90] Government officials in Illinois may have been willing to be patient with Indiana's abatement efforts, but the Chicago media gave heavy coverage to the water pollution issue. On Chicago's south side, the *Daily Calumet* paid special attention to the oil-infested Calumet Park Beach, printing pictures of boat oars coated with oil and slag from steel mills and swimmers covered with oily smudges.[91]

When Illinois governor Otto Kerner, a Democrat, turned to his health director,

Franklin F. Yoder, for advice on the problem, Yoder explained that Secretary Celebrezze had all the information necessary to convene an enforcement conference on the basis of interstate pollution and that it would be wise to let Celebrezze make the decision on a conference.[92] HEW officials demonstrated their usual caution in convening a conference on a heavily industrialized, populated waterway, so Kerner turned to interstate cooperation as a means to demonstrate concern over Lake Michigan. At Kerner's request, in the summer of 1964, Kerner and fellow Democrat Matthew Welsh of Indiana met to discuss the pollution problem in southern Lake Michigan. Subsequently, aides to the two governors met informally to work out an agreement on the issue and came up with a proposal for a formal bi-state compact designed to police interstate pollution involving Indiana and Illinois. Surprisingly, since Indiana had the most to fear from federal intervention, it was the Illinois representatives who pushed for immediate action on the problem. Such a compact would require formal approval by both state legislatures and Congress, but Kerner's aides urged that the two states begin some kind of a joint program in the area before the PHS released its forthcoming report on the Calumet region and southern Lake Michigan. They hoped preemptive joint action by Illinois and Indiana would head off the calling of an enforcement conference.[93]

But before the joint state program got much past the talking stage, Celebrezze finally decided to act and, at the end of 1964, called a conference on interstate water pollution in the area.[94] Chicago Metropolitan Sanitary District officials praised the secretary's decision, citing the danger to the city's water supply from Indiana waste discharges as well as the additional costs of removing pollutants from Chicago's drinking water. District superintendent Vinton Bacon said that "we have been forced to welcome federal intervention even tho [sic] we believe in state's rights and think these problems should be solved locally."[95]

Like other state water pollution control officials, Loring Oeming was concerned about the impact of the growing public uproar over water pollution. Oeming welcomed public support for water pollution control efforts, but he was alarmed at what he viewed as the strident rhetoric, superficial understanding, and unrealistic expectations of the most vocal clean water advocates. These extremists had persuaded many people that the situation was worsening so fast the only solution was to turn responsibility entirely over to the federal government.[96] Critics of the state programs dismissed the views of Oeming and other state regulators. Clean water advocates hoped that the enforcement conferences would usher in a new era of tough enforcement and more stringent waste treatment requirements. But as later events made clear, while federal officials did aim for a higher level of water quality in the region than their state counterparts, in practice federal regulators pursued an enforcement strategy that, like the state approach, was based on consensus decision-making and cooperation.

CHAPTER THREE

Standards and Deadlines

The Changing Nature of Water Pollution Control, 1965–1968

T he years 1965 through 1968 were an important period of transition in the regulation of water pollution in the Great Lakes Basin. Two major developments that emerged during this period would eventually reshape the nature of water pollution control in the region. First, high-ranking federal officials used the enforcement conferences and a new water quality standards program to promote a general upgrade in waste treatment requirements. As industry executives acutely realized, Interior secretary Stewart Udall and officials of the new Federal Water Pollution Control Administration (FWPCA) used their limited regulatory powers and considerable financial inducements to move water pollution control away from the traditional reliance on the "reasonable use" doctrine toward a goal of maximum feasible waste treatment. State officials resisted this shift even as they found themselves drawn along by public opinion and federal authority.

Secretary Udall viewed effective pollution control as an essential part of modern conservation policy, and he became a vigorous advocate of more stringent waste discharge regulation when the national water pollution control program moved to the Department of Interior. At the state level, New York governor Nelson Rockefeller pioneered in the establishment of state funding for sewage treatment plant construction. Rockefeller recognized the large sums that would be needed to install adequate sewage treatment facilities, and his Pure Waters program placed pressure on the federal government and other state governments to increase spending to the levels necessary to meet the challenge.

The second major development in this period concerned the means used to achieve water quality objectives. The government activities stemming from the federal enforcement conferences and the federal water quality standards program initially followed the path of cooperative pragmatism. But by the second half of the 1960s the principles of compromise, consensus, and gradualism standing at the center of this regulatory system came under growing attack from clean water advocates in the region. The federal enforcement conferences were the targets of much of this criticism, since the conference sessions received extensive coverage in the media. Moreover, concerned citizens in these areas had looked to Washington for more effective regulatory action. Many felt betrayed when the conferences failed to produce improvements in water quality. It is not surprising that the first wave of antipollution litigation emanated from Chicago, where the local newspapers made the protection of Lake Michigan into something of a crusade. Thus, by the end of 1968, concerned private citizens and some public officials were losing patience with the established system of regulating water pollution. At the same time, the trend toward maximum feasible waste treatment promised the need for more coercive means of attaining compliance with much greater waste reduction demands.

The Enforcement Conferences Meet

The federal enforcement conferences held in the Great Lakes Basin during 1965 marked an important turning point in the regulation of water pollution control in the region. From this point onward, the federal government would be actively involved in the regulation of area waste dischargers. The heavily publicized conferences demonstrated water pollution's new status as an important public issue attracting more attention from politicians. The highly visible involvement of federal officials in the pollution control process also raised public expectations of immediate improvement in the environment and placed greater pressure on both regulators and dischargers to produce results.

The conference on southern Lake Michigan met in Chicago in March, the Detroit River conference reconvened in Detroit in June, and the Lake Erie conference met in Cleveland and Buffalo in August. The conferees at the Calumet area conference were the federal government, Indiana, Illinois, and the Greater Chicago Metropolitan Sanitary District (MSD). Under Illinois law, the state Sanitary Water Board had no jurisdiction over waste discharges within the sanitary district. Because the sanitary district discharged its wastes to the Illinois River Basin and the discharge from most industries in the district was also drained away from Lake Michigan, the U.S. Steel South Works, which fronted the lakeshore, was the only significant waste source in the sanitary district discharging to the lake. The nonfederal conferees at the Lake Erie conference were Indiana, Michigan, Ohio, Pennsylvania, and New York. Indiana participated in the Lake Erie conference because the Maumee River, a major tributary to the lake, began in Indiana and received waste effluent from municipal and industrial dischargers within the state. The federal government and Michigan were the only two conferees at the Detroit River conference.

About a month before each of the three conferences, the PHS released its report on water quality conditions in the area. The reports confirmed the obvious, that significant pollution existed in these waters. More important, the PHS studies revealed in unprecedented detail the characteristics of the pollution—the major pollutants, their concentrations in different areas, and their impact on the total aquatic environment—and the contributions of the major dischargers. In southern Lake Michigan, industrial dischargers, primarily from Indiana, contributed the bulk of the total effluent load. The PHS pointed to the U.S. Steel operations in Gary and the Inland Steel and Youngstown Sheet and Tube plants in East Chicago as the most significant sources of waste in the area, with oil refineries and chemical plants playing a lesser role. In the Detroit River and adjoining section of Lake Erie, the federal study found that the major sources of waste were the city of Detroit and a number of plants producing steel, chemical, paper, and automobiles. The Lake Erie study dealt with a much larger area than the other two and lacked specific data on individual dischargers but emphasized the problem of eutrophication and the severe pollution existing in the lake's major tributaries and most developed shoreline areas.[1]

The PHS reports also made recommendations for abatement action in the areas studied. Among the most important recommendations were secondary treatment for all municipal treatment plants, general reductions in a number of common industrial pollutants, and regular reporting of industrial effluent data by industry. Because the Detroit River report had been in the making for several years and benefited from the close cooperation of Michigan state officials, it also included recommended effluent reduction levels for individual industrial dischargers. The recommendations for secondary treatment of municipal wastes were aimed in part at combatting the eutrophication problem by reducing nutrient levels, particularly those of phosphorus, which federal scientists believed to be the critical, or controlling, nutrient in the eutrophication process. Federal technicians hoped that effective operation of secondary treatment plants would also reduce phosphorus levels sufficiently so that more expensive nutrient removal measures would not be necessary.[2]

Local newspapers in the Great Lakes region devoted extensive coverage to the PHS reports when they were released and in general played up the importance of the conferences in the weeks before their opening. The figures cited in the PHS reports made good copy. The Detroit River study, for example, found that 1.1 billion gallons of industrial waste were discharged to the river each day, including more than 3,000 gallons of oil, over 80,000 pounds of iron, and more than 8,000 pounds of ammonia. In Cleveland, the *Plain Dealer* ran a series of articles detailing local sources of pollution immediately before the opening of the Lake Erie conference, culminating in a color-photo spread in the newspaper's *Sunday Magazine* depicting graphic scenes of pollution in the Lake Erie Basin. An editorial on the opening day of the conference in the *Plain Dealer* expressed hope that the proceedings would act as a catalyst for accelerated abatement efforts at all levels of government, and pledged that the paper would continue the campaign for a cleanup of Lake Erie.[3]

Given the feelings of state regulators and previous clashes between federal and state officials, observers and some participants were struck at the lack of tension among the conferees and the spirit of cooperation exhibited by the states. A federal representative at the Calumet conference remarked, "I've waited twenty years to see this happen." Even Ohio officials—who, earlier in the year, had all but thrown the federal team out of Youngstown when an enforcement conference had been held regarding the Mahoning River—sought to avoid conflict with Stein and his colleagues. When Stein challenged Emmet Arnold's assertion that "we know what to do about our internal and lakeshore problems in Ohio and are doing it," Arnold sidestepped the issue, clearly wanting to avoid a confrontation. Arnold had little choice, given that Governor Rhodes had opened the conference with a pledge of complete cooperation by state authorities.[4]

Stein was careful to credit the work of the state agencies wherever possible, but he also pressed home the point that the serious pollution problems made federal involvement a necessity. When Pennsylvania secretary of health Charles Wilbar defended the water quality of the Pennsylvania waters in question and criticized the PHS report for inaccuracies, Stein took sharp exception to Wilbar's "rosy and optimistic report," stating that "I have been on three of the rivers here within the past months—the Detroit River, the Cuyahoga and the Buffalo, and if that's the notion of progress being made at the local level in handling the problem, then perhaps the Federal Government is just what is needed in this situation."[5]

The business representatives appearing at the conferences realized that intense media scrutiny and the public outcry over local pollution problems made it imperative that industry position itself firmly on the side of clean water. Hence, there was little talk of the legitimate use of water's waste-assimilative capacity, but great emphasis on past expenditures for pollution control and future programs—either in the planning or in the implementation stage—to reduce the firms' waste contribution even further. The industries stressed the good relationship they enjoyed with state authorities but also expressed a willingness to work with federal officials.

The Detroit River was the one area where industrial dischargers and the federal authorities clashed over policy recommendations. This was not surprising, since the PHS report on the Detroit River area was the only one of the three to include recommendations for reducing the waste constituents of individual dischargers. Detroit area industry representatives, while strongly affirming their support for progress in pollution abatement and willingness to take further steps, questioned many of the effluent reduction objectives recommended for their own companies, claiming these recommendations were unattainable with existing technology or simply impractical. Company officials also questioned just how much water quality improvement such reductions would bring, given the urban-industrial concentration in the area.

Fred Tucker, coordinator of industrial health engineering for National Steel, was one of the most articulate of the industry representatives. Tucker complained that it was unreasonable to expect industry to keep rebuilding and replacing water pollution control equipment to meet constantly changing standards. Tucker made the point that all of National's pollution control facilities were declared acceptable by the state of Michigan at the time of their installation—some only a few years before—but that the new treatment requirements proposed in the PHS report would make most of this new equipment obsolete. When Tucker challenged the practicality of some of the PHS's effluent standards, Hayes Black, industrial waste consultant for the PHS and an acknowledged expert in the field, could reply only that he preferred to think of these values as "objectives" rather than standards. Black explained that in the interest of equity the technical staff had recommended similar reduction requirements for area dischargers, but that "engineering judgement" would have to be used in evaluating each firm's effluent reduction efforts.[6]

The state conferees also invited representatives from interested citizen groups to appear at the conferences. Predictably, the representatives for these organizations emphasized the degraded conditions of the waters in question, urged industries and municipalities to take more action in controlling their wastes, and called for stronger enforcement of antipollution laws. At the Lake Erie enforcement conference in Cleveland, David Blaushild delivered to Stein the more than 200,000 letters and petition signatures gathered during the car dealer's antipollution drive. According to Blaushild, the signatures came from people "who want their God-given heritage of clean waters given back to them—people who cannot, and will not, accept the statements that this shameful degradation of our beaches and lakes and streams is necessary." Blaushild described the impatience that he and other citizens felt with the seemingly endless round of surveys, studies, and meetings that never seemed to accomplish anything. He argued that the tens of thousands of signatures collected on antipollution petitions and the passage by numerous communities of clean water resolutions demonstrated strong, widespread support for regulatory action and disproved the idea that the public was apathetic about water pollution.[7]

The recommendations for remedial abatement measures that emerged from each of the enforcement conferences were ambitious, but quite general in their language. In the case of many of the recommendations, it was left to the state or states to fill in the details concerning specific requirements for dischargers. The individual states also bore the primary responsibility for the actual implementation of most of the recommendations, although Stein made it clear that the federal government would review progress at periodic follow-up sessions of the conference. The final recommendations bore a close resemblance to the recommendations contained in the PHS reports preceding the conferences. The most important recommendations concerned effluent treatment requirements and timetables for construction of needed abatement facilities. The conferees agreed that all municipal waste should receive a minimum of secondary treatment, plus disinfection prior to discharge in most cases. Treatment requirements for industrial dischargers constituted a more complicated matter. In the case of the Detroit River conference, the MWRC agreed to use the PHS report as a guide in setting individual plant effluent treatment requirements. The state conferees at the other meetings lacked such a detailed guide. At the Lake Erie conference, the conferees agreed that industries would have to implement improvements to achieve the maximum reductions of a list of pollutants. At the Calumet conference, the guidelines for industrial waste treatment were even less specific.

The recommendations required the nonfederal conferees to establish treatment facility construction timetables for all municipalities and industries needing remedial measures within six months from the issuance of the formal conference summaries. (At the Detroit River conference the MWRC was originally given a year to establish these schedules, but the deadline was shortened to six months at the Lake Erie conference.) The Detroit River and Calumet area enforcement conferences made no specific recommendations concerning proposed construction timetables, but by the Lake Erie conference in August, federal officials felt confident enough to recommend a construction completion deadline of January 1, 1969, for all dischargers needing remedial measures.

The Calumet area and Lake Erie conferees decided to employ technical advisory committees similar to those working with the Great Lakes–Illinois River Basin Project. The technical committee to the Calumet area conference was to be composed of engineers appointed by each of the conferees and representatives from industry. The committee's task was to formulate water quality criteria for the area and work with the conferees in developing timetables for the construction of needed industrial waste treatment facilities. The Lake Erie conferees decided to establish a similar technical committee, but in this case the committee was to focus on the eutrophication problem and devise recommendations for halting excessive algae growth in the lake.

The PHS's recommendation that Detroit provide secondary treatment for its municipal waste proved the most controversial issue at the round of enforcement conferences. Gerald Remus, general manager of the Detroit water and waste system, argued (as he had at the initial meeting of the conference in 1962) that secondary treatment was not appropriate for the city at this time. Remus believed that secondary treatment was too costly and that there was no certainty about the amount of improvement additional treatment would actually bring in water quality. He maintained that the proper policy for the city was to continue bringing outlying communities on line with the Detroit system, expand storm sewer capacity so as to lessen effluent overflow during heavy rainfall, and gradually upgrade current primary treatment through the use of chemical precipitation. Because of Detroit's intransigence on this issue, the conference's final recommendations omitted any specific reference to secondary treatment and instead stated that all municipalities "be required to provide a degree of treatment sufficient to protect all legitimate uses."[8]

In spite of the general nature of the recommendations and the fact that state agencies were left to fill in the blanks of the most important recommendations, local newspapers portrayed the conferences as "a great victory for the public." Postconference editorials paid little attention to the costs of abatement but urged all parties to follow through on implementing the conferences' recommendations. The intense media coverage and large attendance at the conference hearings helped foster the spirit of cooperation evident throughout most of the proceedings. PHS regional director Wally Poston attributed much of the Calumet area conference's success to the positive support of the Chicago media and their position that "the Federal conference must not fail, if the area is to continue to prosper."[9]

The promises of local newspapers to pay close attention to the implementation of the conference recommendations put officials on notice that their decisions would be subject to close public scrutiny. When George Eagle, Ohio's chief sanitary engineer, briefed members of the Cuyahoga River Basin Water Quality Committee on the Lake Erie enforcement conference, he warned that the state's abatement program for the area would be accelerated "due to the Federal pressure and resulting irresponsible publicity if there are delays." Eagle also said that he believed it was difficult to justify secondary treatment at all municipal plants but explained that "the [Ohio WPC] Board can no longer afford to argue these cases on their merit because of the adverse publicity."[10]

Implementing the Conference Recommendations

With the initial sessions of the federal enforcement conferences covering Lake Erie, the Detroit River, and southern Lake Michigan completed, government regulators sought to turn the ambitious recommendations of these meetings into reality. In spite of their earlier opposition to federal intervention, the state conferees moved quickly to implement the recommendations over which they had control and worked with their federal counterparts where joint action was necessary. Citing the public's heightened concern with water pollution, Loring Oeming advocated fuller cooperation between state and federal officials, reasoning that "whatever can be done to minimize floundering in the glare of this spotlight will certainly be to the advantage of all concerned."[11] Industry officials in the conference areas continued to be on the defensive. Despite mutterings about unrealistic objectives and impossible timetables, plant executives in the end usually pledged their best efforts to meet the new requirements.

In Michigan, the extended PHS investigation of the Detroit River area and the subsequent detailed report enabled the MWRC to move quickly in implementing the conference recommendations. Following the conclusion of the June 1965 federal enforcement conference, Oeming and the other MWRC staff members first determined the uses of the waters in question, relying on a report furnished by the PHS. The Michigan engineers then calculated the water quality levels necessary to protect accepted uses. After comment from PHS staff, area dischargers, and other interested parties, the MWRC adopted ambitious water quality goals that were intended to make the Detroit River suitable for fishing, swimming, and similar recreational uses. The objectives for the Rouge and Raisin Rivers, highly polluted tributaries to the Detroit River, were much lower. As

Oeming noted after a tour of the lifeless and multicolored Rouge, the only way to make that waterway suitable for swimming and fishing would be to eliminate all industry from its banks. After water quality objectives were established, the MWRC staff developed for each significant discharger effluent limitations that were aimed at making the objectives a reality.[12]

Many of the industry representatives appearing at the hearings on the proposed water quality goals and effluent limitations attacked the recommendations as unnecessarily restrictive and claimed that the costs of meeting the objectives would hurt their competitiveness or perhaps force them to curtail production. "They're gasping . . . pleading to save money," scoffed George Harlow, director of the PHS's Detroit River–Lake Erie Project.[13] Members of the MWRC gave more credence to these objections, but after some hesitation they unanimously approved the abatement timetables at a meeting in January 1966 and directed Oeming to prepare stipulations for each discharger based on the approved requirements. The timetables called for the most complex industrial projects to be completed by the end of 1969 and the largest municipal projects by 1970. Smaller projects would be completed earlier.[14]

In a break with the past, once the MWRC decided to proceed with the abatement program, it was unwilling to compromise with dischargers. The commission scheduled its March 1966 meeting to enter into voluntary agreements with municipalities and industries from the Detroit River–Lake Erie area. When more than half of the industries and some municipalities were unwilling to agree to the proposed stipulations without modification, the MWRC rejected most of the polluters' excuses and instructed staff members to initiate formal hearing procedures for the holdouts so they could be placed under orders. Oeming warned the dischargers that if progress was not made, the federal government stood waiting to take enforcement action.[15] The MWRC's firm position brought results: by the end of May, thirty-five of the thirty-six dischargers cited in the federal enforcement conference had agreed to stipulations for abatement.[16]

Gerald Remus, representing the city of Detroit, initially balked at signing the MWRC's stipulation, once again raising questions about the need for secondary treatment and the costs involved. Remus also raised new concerns about the feasibility of the phosphorus removal requirements included in the stipulation. Despite Remus's objections, however, Detroit eventually agreed to the proposed stipulation and its requirements that the city would have the necessary facilities in place to meet the MWRC's treatment requirements by November 1970. Remus, however, made it clear that financing these improvements—estimated to cost more than $100 million—would depend on significant contributions from both the state of Michigan and the federal government. Murray Stein described Detroit's decision to implement secondary treatment as "the biggest victory we have had all year in pollution control."[17]

In southern Lake Michigan, the Technical Advisory Committee appointed by the conferees to the Calumet region federal enforcement conference held a series of meetings in the remaining months of 1965 that were aimed at developing a consensus on water quality standards for the area. The technical committee was composed of eight members from industry and federal, state, and local governments. (The two industry members were from U.S. Steel and the American Oil Company. The city of Chicago and the federal government each contributed two members, while the remaining two slots went to representatives from Indiana and Illinois.) The committee issued its report when the enforcement conference reconvened in January 1966.[18]

The standards recommended by the committee were based on present and future uses of the waters, with a goal of either maintaining present acceptable water quality or improving quality where appropriate. The technical committee applied different standards to different areas. The waters of Lake Michigan were divided into three zones: Open Water, Shore Water, and Inner Harbor Basins. The pollutant limitations in the last zone were less stringent than the other two due to the close proximity of industrial discharge points. Even so, the standards for the Inner Harbor waters were still designed to allow for limited body contact recreation, such as fishing and boating. Stein expressed his pleasure that the members of the technical committee were able to achieve a consensus on such exacting standards.[19] After an executive session, the conferees agreed to adopt the technical committee's standards. Along with these stringent water quality standards, the conferees also adopted an ambitious timetable for the construction of waste treatment facilities that called for all such facilities to be in operation by the end of 1968. Blucher Poole and the members of the Indiana SPCB believed this timetable was unrealistic for the work that needed to be carried out at the large steel mills, but they gave their approval to the timetable, noting the provision that variations from the schedule would be considered on an individual basis.[20]

Like their counterparts in Michigan, industry officials in the Calumet region were alarmed by the pollution abatement demands they now faced. "Unquestionably we're being pushed too fast," one executive complained. The estimated cost to local industry for the abatement program was about $100 million, with much of the needed expenditures concentrated in a handful of firms. Corporate officials' concerns about the tight abatement schedule were accompanied by apprehension about the strict water quality standards adopted by the conference. Industry participation on the technical committee may have continued the form of cooperative pragmatism, but one steel executive explained that business "had little choice in the final decisions. We were clubbed by the antipollution people."[21]

Because of the scope of the Lake Erie federal enforcement conference, the conferees as a group did not establish standards for the American waters of the lake, leaving it up to the individual states to do so under the recently enacted Federal Water Quality Act of 1965. Instead, the individual state boards were to institute steps to ensure that all dischargers to the basin met certain treatment requirements, such as secondary treatment of sewage and "maximum reduction" of enumerated industrial pollutants. Stein and the federal officials were also much more flexible about the schedules for construction of treatment facilities in the Lake Erie conference area than they were in the Calumet region, perhaps because the ills of the former area could be traced to so many more sources.

At the opening sessions of the Lake Erie conference, the federal conferees had recommended January 1, 1969, as the deadline for the completion of both municipal and industrial remedial treatment facilities. But when the conference reconvened in June 1966, all the states except Indiana presented timetables for some larger dischargers that extended beyond the recommended deadline. Nonplussed, Stein countered with a compromise deadline of January 1, 1970, and asked the state conferees to return to the next session with detailed timetables for all dischargers.[22] When the conference finally reconvened in March 1967, the conferees approved the time schedules submitted by each state. Most of the facilities were to be completed during 1968, 1969, and 1970, with a handful of major dischargers—such as the city of Cleveland and Republic Steel's Buffalo plant—scheduled for completion during 1971.[23]

Stewart Udall Takes Over the Federal Program

During the mid-1960s, new laws gave the federal government a greater role in the regulation of water pollution. The Water Quality Act of 1965 gave federal officials the final say on water quality in all interstate waters, and the Clean Water Restoration Act of 1966 greatly increased federal financial support for sewage treatment plant construction. With the transfer of the newly created FWPCA to the Department of Interior in 1966, the federal water quality program gained a powerful new voice in the person of Interior secretary Stewart Udall. As Udall and other federal officials sought to improve the effectiveness of water pollution regulation, they placed special emphasis on the Great Lakes as a top priority area in the national fight to restore the quality of America's waterways.

The creation of the FWPCA recognized the growing importance of water quality protection, but moving the program from the PHS to the new agency created some problems. Since commissioned PHS officials were required to give up their generous uniformed service benefits and assume civil service status when they transferred to the FWPCA, more than a quarter of the middle- and top-level management from the DWSPC chose to either remain in the PHS or retire.[24] To complicate matters, before the FWPCA could become fully operational within the HEW, President Johnson initiated the reorganization in February 1966 that shifted the agency to the Department of Interior. Thus, just as it was ready to begin overseeing the complex process leading to the establishment of interstate water quality standards, the federal program was faced with significant personnel losses and major organizational changes. But for better or worse, much of the program remained intact after the shift. Murray Stein retained his position as chief enforcement officer, while James Quigley assumed the position of FWPCA commissioner. The most obvious difference was the character and style of the man now possessing final responsibility for the federal water pollution control program—Interior Secretary Udall.

The HEW secretaries, with their broad array of responsibilities, had not been strong advocates for pollution control. Udall, by contrast, had fought to acquire the FWPCA for his department, and once the move was made, he devoted much effort to its mission. At the time the FWPCA was created, President Johnson and his advisors were formulating plans to introduce more efficiency and greater coordination into the program by attacking water pollution on a river basin basis instead of artificial state boundaries. Since the Department of Interior had responsibility for many water resource planning and development programs, and Udall had already established himself as a strong public advocate for water pollution control, President Johnson decided to approve the transfer. Opponents of the move raised concerns about Interior's close ties with development interests and lack of experience with water problems in the East, but the president was not swayed.[25]

Stewart Udall, a Mormon from Arizona, began a career as a lawyer before winning a seat in Congress in 1954. After serving three terms and establishing a strong record on conservation issues, Udall became President Kennedy's Interior secretary in 1961, a position he held until the Nixon administration assumed office in 1969. Martin Melosi has described Udall as an important transitional figure in the shift from traditional conservationism to modern environmentalism. Udall had a strong commitment to preserving wilderness areas, protecting wildlife, and other traditional conservation concerns, but he also believed in the need for environmental protection in all parts of the nation. In *The Quiet Crisis,* published in 1963, Udall argued for the need to halt the environmental degradation that was an undesirable side effect of the affluent society.[26]

Udall oversaw the federal water pollution control program for less than three years, but during that time he brought a new sense of urgency to government efforts and embodied the hopes of clean water advocates in the Great Lakes region and throughout the United States. Since Udall assumed authority over an established program operating under well-defined laws, his opportunity for innovation was limited, although he did seek to raise the general level of water quality control through his implementation of the 1965 Water Quality Act and other legislation. Not an outstanding administrator, Udall probably made his greatest mark on water pollution control through his efforts to educate the public and dramatize the need for tough regulation of polluters. The seemingly tireless Udall—with his crew cut, strong oratory, and penchant for the "strenuous life"—seemed a throwback to the heroic age of conservation. During his last years as Interior secretary, he made frequent appearances in the Great Lakes region to take part in conferences and promote stronger pollution control.[27]

Udall and other top federal officials gave the Great Lakes their highest priority. Appearing at the opening of the June 1966 session of the Lake Erie enforcement conference in Cleveland, Udall described the Great Lakes region as "one of the real battlegrounds or proving grounds with regard to the war on water pollution." Udall said that if the authorities could solve the problem in Lake Erie, then they could certainly get the job done anywhere else in the nation.[28] Two months later, President Johnson took a Coast Guard cutter tour of Buffalo

Harbor with Governor Rockefeller to observe the area's pollution problems first-hand. Local pollution control activist Stanley Spisiak used the occasion to show Johnson a bucket of Buffalo River bottom sludge and a half-dozen small glass jars containing multicolored water from the river. In prepared remarks, Johnson declared that "Lake Erie must be saved." Echoing a theme frequently expressed by Udall, Johnson said that his administration's war on pollution would clean up the problems that America's abundance had created.[29]

By the end of 1965, many federal officials had come to believe that the enforcement powers of the FWPCA needed to be strengthened and broadened if significant progress was to be made in cleaning up the nation's waterways. When President Johnson signed the Water Quality Act of 1965 into law, he used the occasion to publicly condemn paper mills, chemical plants, and oil refineries for their pollution of the nation's waterways and promised "additional, bolder legislation" in the years to come. That same month, an interagency federal task force created by Johnson to study pollution abatement issued a report that criticized the cumbersome enforcement procedures available under current federal water pollution laws. The task force recommended a variety of changes to simplify and enhance federal enforcement powers.[30]

In March 1966, Johnson included in his special message on conservation and natural resources a number of proposed amendments to the Federal Water Pollution Control Act that followed the task force's lead. Among other things, the amendments would have extended federal oversight of state water quality standards to intrastate streams and eliminated mandatory delays before federal court action could begin. But the liberal Democrats in Congress who usually played the leading role in designing water quality legislation rejected the administration's proposals for greater federal enforcement powers. Instead, Senator Edmund Muskie and his colleagues urged that the new water quality standards program be given a chance to work. They prepared legislation aimed at dramatically increasing the amount of federal aid available for sewage treatment plant construction.[31]

Setting Water Quality Standards:
The Movement away from Reasonable Use

The process of setting interstate water quality standards forced government officials to examine and clarify on a public stage their assumptions about water pollution control and their goals for the future. Stewart Udall and officials in the Department of Interior and the FWPCA used their position as final decision-makers on interstate water quality standards to promote a general increase in waste treatment levels. The system of assigning various grades of quality to different waters based on their use and other local factors only made explicit the procedure that most states had been following for years. By this time, however, state officials in the Great Lakes Basin were willing to give a higher priority to recreation and aesthetic values. Even more important, the state regulators were now operating with a federal Big Brother looking over their shoulder. The clean water advocates who appeared at the public hearings on the state water quality standards condemned the practice of "water quality zoning" and called for a new policy requiring maximum feasible treatment. The experienced engineers from industry and state government paid little heed to the housewives and sportsmen who made these arguments, but statements by high-ranking federal officials indicated that the latter group's views represented the future of water quality control, at least as the federal government envisioned it.

The Water Quality Act of 1965, like most legislation, was a product of compromise. This law grew out of widespread dissatisfaction in Congress with the slow pace of water pollution abatement. Senator Muskie and some other members of Congress originally hoped to provide the secretary of HEW with the authority to set both water quality and effluent standards for all navigable waters within the United States, but this proposal was too drastic for the majority of legislators. Under the final legislation, each state was to establish water quality standards for all of its interstate waters by June 30, 1967. The standards were subject to the approval of the secretary of HEW, who was also given the power to set standards for the states if they did not establish them.[32]

Under the new law, a state's standards consisted of two main elements: (1) water quality criteria for each major use, such as drinking water or recreation, and designation of the waters in which they were applicable; and (2) a formal plan for implementing and enforcing the water criteria. When the FWPCA joined the Interior in May 1966, the former issued guidelines for the establishment of the interstate water quality standards. According to the Water Quality Act, the standards were intended to protect clean waters and enhance water quality in degraded areas. The guidelines explained that, ideally, the standards established in the degraded waterways would be higher than present water quality; in no case would standards lower than present water quality be accepted. The application of this guideline to relatively pure, undeveloped waters was more complex. At first, the FWPCA was willing to allow degradation of these waters as long as existing water uses were protected, but pressure from conservation organizations later forced the agency to adopt a more rigid stance. Another important—and controversial—guideline in effect required that all municipalities and industries discharging to interstate waters provide secondary waste treatment or its industrial equivalent.[33]

Federal officials made it clear that if the states did not do an effective job in setting water quality standards, they faced an even greater federal role in pollution control policy. FWPCA Commissioner Quigley once again threatened his state counterparts with displacement, warning that the standards program was "the states' last big opportunity to do their part in this job. . . . If the states fail, the trophy is going to the Federal government by default."[34] The new water quality standards were intended to simplify enforcement; under this new approach, the federal government could enforce the water quality standards by bringing suit in federal court against the polluters responsible for standards violations. The HEW secretary was, however, still required to give the polluter notice and then a 180-day waiting period in which to rectify the problem and meet the standards. And, as in the enforcement conference process, the consent of the state governor was required before legal action could be taken in cases involving standards violations that were intrastate in nature.

The water pollution control agencies in the Great Lakes states followed similar procedures in setting water quality standards. First, they established water quality criteria for each basic water use. This meant setting numerical limits on individual pollutants and establishing minimum levels of dissolved oxygen. Wisconsin, for example, mandated that waters designated for swimming should not exceed an average coliform count of 1,000 per 100 milliliters, and a maximum of 2,500 per 100 milliliters. State officials assigned criteria to interstate waters based on present and desired uses.[35] The state agencies then developed an implementation and enforcement plan for each water basin or distinct area of water that was designed to protect or attain the assigned uses. The keystone of these plans was the individual abatement schedules for those dischargers requiring a higher degree of waste reduction in their effluent. Once a plan was completed, the state conducted a hearing in the affected area to allow interested parties and individuals an opportunity to comment on the proposed criteria and implementation plan. After the hearing, state officials made any adjustments they believed were warranted. The final, informal step before the state submitted its proposed water quality standards to Washington was a series of meetings between state and FW-PCA officials to reconcile any major differences.

Under the plans established by the Great Lakes states, the open waters of the lakes and connecting channels were to be acceptable for all water uses, whereas harbor areas and certain shoreline areas adjacent to major effluent discharge areas were not required to meet the more stringent criteria. The criteria adopted for interstate tributaries to the lakes were based on physical characteristics, current water quality and uses, and other local factors.[36] Industry supported the concept of basing water quality requirements on primary use, but business officials in the Great Lakes Basin were alarmed at the exacting standards proposed by state officials, even in highly industrialized areas.

At the MWRC's hearing on water quality standards for the Detroit River, a horrified Fred Tucker of National Steel observed that the recommended criteria were in many cases even more restrictive than the water quality objectives adopted by the MWRC after the Detroit River federal enforcement conference. In northwestern Indiana, the SPCB adopted the standards approved by the Calumet region federal enforcement conference when establishing criteria for the area. The enforcement conference, however, had failed to establish criteria for the Indiana Harbor Ship Canal, which carried the wastes from Inland Steel, Youngstown Sheet and Tube, and a number of other industries into Lake Michigan. When the board held its hearing on the Lake Michigan Basin in April 1967, representatives from individual firms and local chambers of commerce argued that some of the recommended criteria were not feasible from either a technical or an economic standpoint, especially those proposed for the ship canal. The ship canal criteria, in their eyes, would provide a level of quality not necessary for a body of water used mainly as a waste conduit.[37]

Representatives from conservation and civic organizations questioned the very concept of assigning levels of water quality to different waters based on use and called for a single standard of high quality for all waters, at least as an objective. Representatives from the Izaak Walton League were especially prominent at the state hearings on water quality standards held in the Great Lakes Basin. Arnold Reitze, a league member and law professor at Cleveland's Case Western Reserve University, was one of the most articulate critics of the water quality standards program. Reitze argued that the concept of multiple water quality standards was essentially undemocratic because it allowed public water resources to be of low quality in order to benefit a small number of people. According to Reitze, "a system of water zoning has been created that legitimizes continued low-quality water."[38]

Clean water advocates also expressed concern about the protection of high quality waters in undeveloped areas. They argued that the practice of allowing pure waters to decline to the levels of established use criteria clashed with the philosophy of water quality enhancement that was supposedly the guiding principle of the Water Quality Act. When the Department of Interior began accepting state standards that allowed for such degradation, the outcry from local and national conservation groups and some members of Congress forced Secretary Udall to modify the administration's policy on this issue. All states would now be required to include a "nondegradation statement" in their standards, declaring that interstate waters currently exceeding the established water quality standards would not be allowed to decline except in cases where "such change is justifiable as a result of necessary economic or social development."[39] The nondegradation issue generated much controversy, but in practice the Department of Interior proved willing to accept state versions of the nondegradation statement that changed the wording so as to undermine the original intent or that added qualifications reaffirming the state's final authority in the matter. By the beginning of 1969, the Department of Interior had approved forty-eight of the fifty state submissions, although many were approved with exceptions such as temperature standards, pending further negotiation.[40]

At this point, federal officials were more concerned with raising the general level of waste treatment in the United States than with preserving the nation's highest quality waters in a pristine state. While intrastate waters had been omitted from the standards-setting requirements of the Water Quality Act of 1965, a provision in the Clean Water Restoration Act of 1966 provided financial incentives for those states that established acceptable intrastate water quality standards. As a result, the vast majority of states adopted a system of standards applicable to all surface waters within their boundaries.[41]

State officials in the Great Lakes region and other parts of the United States sensed the importance of the changes emanating from Washington. Clarence Klassen, technical secretary of the Illinois Sanitary Water Board since 1935, addressed these changes when he spoke before the annual meeting of the WPCF in October 1967. Klassen noted a major shift in the federal approach to pollution control following the Water Quality Act of 1965 and the creation of the FWPCA: "Suddenly, the conservationist, the recreationist, the purist burst forth to dominate the scene and to take over to a large extent the administrative control of the federal program." Klassen worried that the ambitious objectives for water quality implicit in the federal standards-setting guidelines might appeal to the public but would prove unattainable—and unenforceable—in many areas.[42] A week later, municipal and state engineers joined with their counterparts from industry at the water resources engineering meeting of the American Society of Civil Engineers to attack the federal water pollution control program. At the keynote session of the meeting—"What Price Clean Waters?"—panelists argued that the massive public expenditures necessary to attain pristine pure water could be better used for other pressing social needs. State and industry engineers also questioned the rationale for mandating at least secondary treatment for all interstate waters.[43]

The following summer, the recently appointed commissioner of the FWPCA, Joe Moore Jr., responded to these criticisms in an article in *Civil Engineering*. Moore admitted that there had been a shift in federal policy from "reasonable use" and conscious utilization of stream-assimilative capacity to maximum feasible waste treatment. He attributed the change to a number of factors, including continued population and industrial growth, concerns about lake eutrophication, a greater interest in the aesthetic value of clean water, and the growing political strength and effectiveness of conservation groups and their allies. Moore defended a general requirement for secondary treatment or its equivalent on the grounds of need and also economic and technical feasibility. In addition, he noted that, in many areas, additional types of treatment such as phosphorus removal were necessary if acceptable water quality was to be preserved or restored.[44]

Despite their concerns about cost, some sanitary engineers in the Great Lakes states were also expressing doubts about the continued viability of the "reasonable use" doctrine. Speaking at the June 1966 session of the Lake Erie enforcement conference, Robert Hennigan of the New York health department called on those involved in water quality management to adjust their thinking to present realities and move beyond the outmoded approach of linking waste treatment requirements to stream-assimilative capacity. Hennigan argued that "the waste assimilation capacities of our lakes and streams has [*sic*] been overworked to the point that the very concept has played a major role in the deterioration of water quality across the country."[45] Even Klassen found himself adopting a new attitude about waste treatment, at least for public consumption. When Richard Billings of Kimberly-Clark Corporation appeared at the opening session of the four-state Lake Michigan federal enforcement conference in February 1968 to emphasize the importance of cost-benefit analysis and the need to avoid "treatment for treatment's sake," Klassen took him to task for following "the old fashioned approach" to pollution abatement. The sizable audience applauded Klassen's remarks.[46]

Paying for Municipal Pollution Abatement

Higher expectations about water quality would not be enough, of course. The costs to implement the needed improvements in water quality treatment were staggering. In February 1966, regional program director Wally Poston of the FW-PCA estimated that a ten-year program to clean up all the Great Lakes would cost about $20 billion.[47] New York governor Nelson Rockefeller was the first governor in the region to face up to the costs involved and take dramatic action. Rockefeller first made water pollution control a significant part of his administration's agenda in 1962, when he supported bills to provide limited state funding for municipal sewage treatment plant planning, construction, and operation. The bills became laws that same year and appeared to herald a new state commitment to pollution abatement, but the more expensive construction and operating grant programs fell victim to the governor's budget battles with the legislature. The bottom line was that, in budget considerations at this point, water pollution control still took a back seat to more traditional concerns such as education.[48]

The political climate was far different at the end of 1964, however, when Rockefeller announced his bold proposal to clean up New York's waters in six years. For Rockefeller, the basic problem in water pollution abatement was money. His seven-point Pure Waters program was aimed at stimulating the construction of needed treatment facilities through an ambitious grant program for municipalities and tax incentives for industry. To meet New York's backlog of existing treatment needs and to deal with new needs, Rockefeller proposed total expenditures of over $1.7 billion. A billion-dollar bond issue, subject to the approval of state voters, would allow the state of New York to assume 30 percent of the construction costs of each municipal sewage treatment plant. Under this plan, the federal government would fund another 30 percent of each plant's costs, while the local government or governments would cover the remaining 40 percent.[49]

A major problem was that at this time the federal government's grant program had a $600,000 limit on individual projects and a $2.4 million limit on joint projects. In addition, under the federal law, 50 percent of the total grant money was to be directed to communities with under 125,000 people. Rockefeller charged that these provisions discriminated against large urban states like New York. Acting on the assumption that the magnitude of the problem would eventually force Congress to drop these limitations and raise total spending on grants, Rockefeller decided to "prefinance" the federal government's share of projects, when necessary. In other words, New York would pay Washington's 30 percent share at the beginning of a project, with the expectation that the federal government would eventually reimburse New York.[50] The Pure Waters program also included tougher enforcement of state pollution laws and tax incentives designed to spur construction of new industrial treatment facilities. But the main thrust of the program was providing funds for the construction of municipal treatment facilities.[51]

The ease with which the Pure Waters program passed through the New York legislature and the overwhelming passage of the bond issue in November 1965 demonstrated the political appeal of water pollution control. In this case, Rockefeller's proposal also benefited from the impact of the severe drought that hit the Northeast in the first half of the 1960s. The drought raised public awareness about the importance of water availability, and the inability of local governments to use many rivers because they were so polluted vividly demonstrated the link between water quantity and water quality. The bills comprising the Pure Waters program passed both houses of the New York legislature unanimously after being sponsored jointly by the majority and minority leaders of each body. Even more important, the $1 billion bond issue won approval at the polls by a vote of 4 to 1.[52]

While Rockefeller was willing to spend large sums of state money on water pollution abatement and advocate increased federal funding, he continued to oppose federal intervention in the regulation of waste dischargers, arguing that the states should retain primary authority for the control of water pollution within their borders. At the Buffalo meeting of the Lake Erie conference, Rockefeller turned down a request from Senator Robert Kennedy and Representative Richard McCarthy that he ask the federal government to extend the conference to cover New York's intrastate pollution problems in the Niagara River and Lake Ontario. Rockefeller stated that New York would gain no real benefit from such a move because it would not result in any more financial aid from Washington. He argued that because of New York's strong state program, any federal enforcement activity would only duplicate and hinder state action.[53]

Nelson Rockefeller's Pure Waters program was important for two reasons. First, the large sums involved and Rockefeller's public pronouncements on the spending provisions helped dramatize the inadequacy of the existing federal grant program. Second, the provisions for state financing of municipal treatment plants served as a model for other states interested in taking further steps to combat pollution. Federal officials such as Interior secretary Stewart Udall applauded Rockefeller's program and urged other states to follow New York's lead.[54] The Pure Waters program recognized the financial limitations that many cities faced. Larger cities, especially, faced with myriad urban problems, had to turn to their state capitols and Washington for aid. The Clean Water Restoration Act of 1966 promised the needed support, but like the rest of the Great Society programs, actual appropriations failed to match initial promises.

The authors of the Clean Water Restoration Act largely disregarded the Johnson administration's proposals for promoting a river basin approach to pollution abatement and instead dramatically increased federal funding for the construction of municipal waste treatment plants. The Water Quality Act of 1965 had raised the yearly federal authorization for the construction grant program from $100 to $150 million for fiscal years 1966 and 1967. The 1966 legislation went far beyond this, authorizing expenditures of $450 million in fiscal year 1968,

$700 million in fiscal 1969, $1 billion in fiscal 1970, and $1.25 billion in fiscal 1971.[55] Just as important, the Clean Water Restoration Act removed the dollar ceilings on individual projects and provided for a federal contribution of 50 percent to individual projects in states that agreed to pay 25 percent of the cost and establish enforceable water quality standards for the project's receiving waters. This last provision was designed to encourage states to set up their own grant programs and establish intrastate water quality standards, and it succeeded on both counts. Finally, in large part because of lobbying by Governor Rockefeller, the 1966 act allowed for the reimbursement of states or municipalities that financed the construction of approved treatment projects with no federal aid or less than the approved amount. This provision, however, was not to be construed as a commitment by the national government to pay for any part of the approved project if funds were not available.[56]

By the end of 1968, all the Great Lakes states except Minnesota and Pennsylvania had established state sewage treatment plant construction grant programs to supplement the federal contribution, although these varied in size and scope. The enormous costs involved sometimes necessitated large bond issues that required approval at the polls. Even with rising public concern about the environment, passage was not guaranteed. In the November 1968 election, Michigan voters approved a $335 million state bond issue to fund sewage treatment plant construction by more than a two-to-one margin.[57] In Illinois, however, a $1 billion natural resources bond issue that included $250 million for sewage treatment plant construction went down to defeat.[58]

State and local officials in the Great Lakes states were obviously pleased with the tremendous increases in expenditures promised by the Clean Waters Restoration Act, but their pleasure turned to disillusionment as Congress and the Johnson administration failed to follow through on the promised financial support. As the costs of fighting the Vietnam War escalated, the water pollution control program (like other domestic programs) was forced to make do with lower levels of funding than had been planned. The federal law authorized $450 million for fiscal year 1968 construction grants, but only $203 million was actually appropriated. The gap for fiscal 1969 was even more severe—$700 million authorized and $214 million appropriated. Murray Stein and other federal officials defended the administration from critics who complained about budget cuts by pointing out that actual spending on the grant program had increased and that it was not unusual for program appropriations to fall short of authorized spending levels.[59]

Still, in order to comply with federal enforcement conference recommendations and new water quality standards, municipalities faced large expenditures for upgrading their treatment facilities by specific deadlines. According to Loring Oeming, the failure of Congress to appropriate the full amount of money authorized for the construction grant program slowed down local cleanup efforts, even in the case of municipalities that could probably afford to finance the entire project themselves. City officials were understandably unwilling to proceed with entirely local financing when there was a possibility that Lansing and Washington would pick up 75 percent of the tab. But since only a minority was able to obtain the grants each year, given current federal expenditures, the rest put off action, hoping to obtain the grants in future years.[60]

The ability of the federal government to deliver on its financial promises was especially critical to large cities such as Detroit and Cleveland. Detroit's Remus made it clear when the city entered into a stipulation with the MWRC that attainment of the proposed treatment levels would depend on both state and federal financing. In Cleveland, a study by engineering consultants estimated that it would cost over $200 million to make the needed improvements in the city's sewage treatment system. Cleveland's new mayor Carl Stokes announced in the summer of 1968 that his administration planned to present a $100 million bond issue to voters to help pay for these improvements. Cleveland voters that November approved the bond issue (the largest ever passed for water pollution control at the local level) by a two-to-one margin. But as Stokes and other city officials made clear, they were counting on matching funds from the federal and state governments to make up the difference.[61]

The improvement of municipal treatment plants in the Great Lakes Basin was critical for more than just controlling sewage waste. Industrial wastes discharged to municipal treatment systems often made up a significant component of the total waste load. In Milwaukee, for example, industrial effluent channeled to the city's waste treatment plants represented a BOD population equivalent of 1,570,000, as compared to the actual population of 1,080,000 served by these facilities.[62] In fact, during this period industrial discharges were gradually accounting for a larger percentage of the total wastes handled by municipal plants. President Johnson's special task force on pollution abatement recommended in its October 1965 report that the federal government encourage the treatment of industrial wastes in municipal plants, and federal officials subsequently adopted this line.[63] Municipal treatment was an attractive option for many industries, especially smaller operations, because it avoided large capital costs, maintenance headaches, and direct government regulation. But the large volume and unique characteristics of industrial effluent sometimes acted to lower the operating efficiency of municipal plants designed to handle sewage.[64]

Dealing with Lake Eutrophication

The presence of large amounts of phosphorus in municipal sewage also presented government regulators with a difficult issue. The most complex technical problem that the Lake Erie conferees had to deal with was the eutrophication of the lake. Further technical evidence that nutrient over-enrichment was contributing to the degradation of Lake Ontario and Lake Michigan made the need for a feasible solution to this problem even more urgent. By the summer of 1966, Wally Poston, FWPCA regional director, was pointing to accelerated eutrophication as the major problem in the Lake Ontario Basin. In Lake Michigan, the increasing presence of stinking piles of *Cladaphora* on the lake's southern beaches indicated that Lake Michigan faced a similar problem. In September 1967, a federal official estimated that the algae problem in Lake Michigan was similar to conditions that had existed in Lake Erie and Lake Ontario in the 1930s. Immediate action was needed if Lake Michigan was to avoid following the downward path of its sister lakes.[65]

At the June 1966 meeting of the Lake Erie enforcement conference, the technical committee appointed to study the eutrophication problem presented an interim report. The committee—composed of a representative from each state and the federal government—acknowledged gaps in understanding of the eutrophication process but concluded that phosphorus reduction was the best approach to the problem. Other contributing nutrients entered the lake through naturally occuring processes that would be impossible to control. It was the high concentrations of phosphorus created by human activity that caused the excessive growth in algae. The technical committee urged more work on developing means to improve phosphorus removal at municipal treatment plants, since municipal effluent was the major source of this nutrient.[66]

At the initial meeting of the Lake Erie enforcement conference, the conferees had hoped to achieve significant nutrient removal at municipal treatment plants through the efficient operation of secondary treatment facilities. Unfortunately, this approach had not been very fruitful. At the same session that reviewed the technical committee's report, Indiana reported that the Fort Wayne secondary treatment plant had been experimenting with various alterations to standard operating procedures in the hopes of attaining high phosphorus removal, but so far 20 percent had been the maximum reduction obtained. At Erie, Pennsylvania, similar experiments resulted in phosphorus reduction that varied between 7 and 27 percent. Since the phosphates contained in synthetic detergents accounted for a large amount of the phosphorus in municipal effluent, some concerned parties argued that the elimination or at least reduction of the phosphate content in detergents was a logical step in battling lake eutrophication.[67]

On October 13, 1966, Charles Bueltman, technical director of the Soap and Detergent Association, met with the Lake Erie Technical Committee to discuss the eutrophication problem in Lake Erie and the role of detergent phosphates. Bueltman emphasized the difficulty of finding a substitute for detergent phosphates that would meet all the safety and performance requirements. When asked about federal research assistance, Bueltman said that the industry's current research effort in this area was adequate and that federal assistance or intervention could prove detrimental to the task. In spite of Bueltman's views, the technical committee recommended in its next report that the detergent industry and the federal government "should promote and encourage the research and development of a suitable substitute."[68]

In July 1967, Secretary Udall and other senior Interior officials met with top executives from the soap and detergent industry to discuss the role of detergents in the eutrophication problem. The result of this meeting was the creation of the

Joint Industry-Government Task Force on Eutrophication. Charles Bueltman served as the group's chair, which included representatives from both government and industry. This body was firmly rooted in the twentieth-century form of American corporatism that emphasized business-government interaction and the resolution of economic and social problems through informal cooperation and the joint application of technical expertise.[69] The Interior press release announcing the formation of the task force, however, indicated that the two parties to the agreement may have possessed different perceptions about the best means of approaching the problem. The press release said that the task force would consider the role of detergent phosphates in eutrophication and possible substitutes, but it also noted that "the industry representatives stated their considered opinion that the likelihood of finding a practical solution to eutrophication is greatest if an over-all research approach to the problem of eutrophication is pursued."[70]

By early 1968, experimental work conducted under the auspices of the FWPCA had resulted in the development of feasible methods for achieving significant phosphorus reduction at wastewater treatment plants. The most practical method was lime precipitation. After waste effluent received biological (secondary) treatment, lime was added to the effluent. This caused the phosphorus in the effluent to flocculate so it could be separated from the waste stream. The capital investment for this additional operation was relatively small. The daily operating costs of the treatment plant, however, could be expected to increase by at least 50 percent, a sizable amount at larger treatment plants. For the city of Cleveland, this translated into approximately $10,000 per day of additional operating costs or about $4 million a year.[71]

The conferees representing the federal government, Michigan, Indiana, and Pennsylvania reached agreement that municipalities in the conference area should be required to reduce phosphorus levels in their effluent by 80 percent, but the conferees from New York and Ohio balked at this requirement, citing cost concerns and the need to upgrade primary treatment facilities first. Ohio officials were particularly reluctant to adopt this target. Since Ohio had many more municipal treatment plants in the Lake Erie Basin than any other state, Ohio's bill for phosphorus reduction would dwarf the others.[72]

The Ohio sanitary engineers relied on the methods of cooperative pragmatism to attack this problem, calling on the advice and experience of industry, municipalities, and other interested parties. The resulting Advisory Committee on Algae Growth developed a series of questions for presentation at the upcoming technical session of the enforcement conference. The committee questioned whether the relationship between phosphorus and eutrophication was clear enough to justify the proposed expenditures. An FWPCA official who sat in on one of the committee's meetings noted cynically that this stance was not surprising, given the presence on the committee of representatives from Procter and Gamble, consulting engineers, and sewage treatment plant operators.[73] At the August 1968 technical session, the FWPCA's scientists defended the need for phosphorus removal if the eutrophication problem was to be brought under control. In fact, the latest studies indicated that even 80 percent phosphorus removal would not be sufficient to bring the lake phosphorus loading down to desired levels, given projected increases in agricultural runoff and municipal discharges. At that time, however, under existing technology, an 80 percent removal rate was the most practical target.[74]

At the follow-up session of the conference in October, Murray Stein was finally able to attain a consensus on phosphorus reduction. In large part because Ohio objected to adopting a policy of 80 percent phosphorus removal for all dischargers, the conferees agreed to an objective requiring each state to achieve an 80 percent reduction in its total phosphorus input, with the understanding that the state could grant variances to individual sources—especially small dischargers—so long as the total phosphorus reduction goal was attained.[75] For many people involved in pollution issues in the Great Lakes region, the high costs and technical challenges of removing phosphorus from municipal sewage pointed to the need to eliminate the greatest source of phosphorus in municipal effluent—detergent phosphates.

The Lake Michigan Enforcement Conference

Even though the federal enforcement conferences failed to generate immediate results, clean water advocates in the Great Lakes states still looked to Washington for greater abatement progress. Moreover, growing popular awareness about the widespread nature of the eutrophication problem supported the position of those who argued that meaningful water pollution control had to be interstate in nature. The by now familiar debate over the need for federal involvement was played out again after 1965—this time concerning all of Lake Michigan. When the four-state federal enforcement conference on the pollution of Lake Michigan finally convened early in 1968, the conferees clashed over the need for uniform waste treatment requirements and addressed a number of relatively new pollution problems that had been overlooked in the past.

As further studies revealed the deep-seated problems facing Lake Michigan, observers such as the *Chicago Tribune* and some government officials began to question the wisdom of confining federal enforcement action to the southern end of the lake. Senator Gaylord Nelson of Wisconsin was an early and vocal proponent of expanding the federal enforcement conference to cover all of Lake Michigan. At the February 1966 technical session of the Calumet area federal enforcement conference, Murray Stein mentioned the polluted condition of the lake waters near Milwaukee and stated the willingness of federal officials to help with the situation—if they were invited in.[76] Stein's remarks set off a chain of events familiar to anyone who had observed the political maneuvering that preceded the convening of the earlier federal enforcement conferences in the Great Lakes Basin. The following month, Senator Nelson sent letters to Governor Warren Knowles of Wisconsin, Governor George Romney of Michigan, and Governor

Karl Rolvaag of Minnesota recommending that they request a federal interstate enforcement conference for Lake Superior and the western shore of Lake Michigan. Knowles and Romney were Republicans, Rolvaag a Democrat. Knowles responded by raising the possibility of a tri-state governors' meeting on the issue, but Romney advised against such a meeting, arguing that it might imply the need for federal intervention in these waters.[77]

Governor Warren Knowles, like his counterparts in Michigan and New York, was a moderate Republican with a sincere interest in protecting the environment through progressive state programs. Knowles, first elected governor in 1964 and reelected twice, was an avid sportsman who had long been active in Wisconsin conservation circles. Knowles had close ties to the Wisconsin paper industry, but he argued it was in the best interests of the business community to recognize that changing social attitudes made it necessary for industry to make greater efforts in this area than it had in the past. The issue was not going to go away; if significant strides forward were not made by responsible leaders like himself, they would leave the field open to politicians who had little sympathy for industry's concerns. Although Knowles was committed to environmental reform, he was put off by the exaggerated rhetoric of politicians who portrayed the state's waterways as running sewers and cesspools. He condemned the "shear [sic] demagoguery . . . practiced by Nelson, [Congressman Henry] Reuss, and others who profess interest, but only for political purposes."[78]

Nelson continued to pressure Knowles to request a federal enforcement conference. Shortly after his letter to the governors, Nelson met with Murray Stein in Washington and then emerged to make a statement that federal pollution experts were ready to swing into action to solve water quality problems on Wisconsin's side of Lake Michigan—if only Governor Knowles would request their aid. Stein appeared in Wisconsin several weeks later, where he stated publicly that Wisconsin was losing its position as a leader in pollution control and that Governor Knowles was blocking needed federal support. At this point Knowles had had enough, and he fired off an angry letter to President Johnson, HEW secretary John Gardner, and Wisconsin members of Congress defending his record and criticizing Stein for making unwarranted public statements.[79]

With Knowles up for reelection in November 1966, the issue of water pollution and federal intervention became an important issue in Wisconsin politics. Beginning in 1964, state authorities and the paper industry came under growing criticism for their failure to deal with Wisconsin's water pollution problems. That year, the Wisconsin Resource Conservation Council published a study—based on data obtained from the State Committee on Water Pollution—that charged the paper industry with contributing 78 percent of the total BOD discharged to the state's waters. The report criticized the industry for taking insufficient steps to abate its wastes and blamed the Committee on Water Pollution for almost ignoring the paper mills and concentrating its efforts on small dairy and canning plants. The report implied that paper industry interests had effectively "captured" the state's pollution control apparatus, in part through the influence of the paper-dominated Advisory Committee on Waste Disposal, which met regularly with state officials to "exchange views."[80]

Newspapers with strong Democratic sympathies now criticized Knowles for failing to request a federal enforcement conference, charging that the governor and his industrial advisors were most concerned with avoiding strong enforcement action that could prove costly to the Wisconsin paper industry. Governor Knowles's opponent in the upcoming election, Lieutenant Governor Patrick Lucey, sought to capitalize on the issue by pledging to request such a conference as one of his first official acts as governor. Knowles moved to undercut his critics with strong action to revitalize the state's water pollution control program. The Wisconsin Water Resources Act of 1966, signed by the governor in June 1966, completely reorganized the state's water pollution control program, increasing its budget, size, and powers. Secretary Udall praised the legislation as the best in the country.[81] In June, Knowles also held two sessions of a governor's conference on Lake Michigan pollution to demonstrate his concern for the local water quality. The federal and state officials who participated in the conference agreed on the existence of severe pollution in the Wisconsin waters of Lake Michigan, but the former emphasized federal solutions to the problem, while Wisconsin officials stressed state powers and responsibilities.[82] Senator Nelson and others continued to press for a federal enforcement conference on all of Lake Michigan, but now the focus of controversy moved back to Chicago.

In the summer of 1967, a massive die-off of alewives inundated the beaches of Lake Michigan, generating an awful stench and creating a public nuisance. Water pollution did not appear to play a direct role in the deaths, but the alewife catastrophe riveted public attention on the problems of the lake. Later that summer, a huge bloom of *Cladaphora,* unprecedented in size, coated the lakeshore and dramatized the eutrophication problem. The *Tribune* and the other Chicago papers covered these events closely.[83]

While state regulatory officials in Michigan and Wisconsin maintained that a four-state federal enforcement conference on the lake's problems would be premature, their counterparts in Illinois and Indiana were more ambivalent. The Calumet Region conference had already involved federal officials directly in the latter states' Lake Michigan abatement programs. Klassen believed that the other two lake states needed to be involved, or else it would appear that Illinois and Indiana were largely responsible for the region's pollution problem, which was far from the case. In somewhat indirect language, Klassen pressed the federal conferees to call a four-state conference, but Stein made it clear that Secretary Udall and the FWPCA would prefer to have one of the governors request the conference. Blucher Poole of Indiana wanted to see the other two states involved, but he did not want to expand the conference.[84]

By fall 1967, the well-publicized symptoms of the continuing decline in southern Lake Michigan water quality—including intense media coverage of a massive oil slick emanating from Indiana Harbor—had convinced many people in the Chicago area that a federal enforcement conference dealing with the entire

Lake Michigan Basin was necessary. In a telegram to Secretary Udall dated September 22, 1967, the board of trustees of the Chicago MSD asked that the Calumet region enforcement conference be expanded to include all four states bordering Lake Michigan. A week later the *Chicago Tribune,* despite the newspaper's traditional opposition to federal power, also called for the convening of a four-state conference. Governor Otto Kerner of Illinois, like Knowles before him, sought to defuse the issue by arranging a meeting of the four Lake Michigan governors, but he was no more successful in this regard than Knowles had been, and the meeting never came off. On November 22, 1967, Kerner wired Udall with a request that the Interior secretary convene an enforcement conference on Lake Michigan.[85]

The Lake Michigan enforcement conference opened in Chicago on January 31, 1968, with over six hundred people in attendance. Illness prevented Secretary Udall from attending, but in a written statement read to the audience by Interior assistant secretary Max Edwards, Udall made a plea for decisive action by the conferees. "Delay means death to Lake Michigan," Udall warned. The Interior secretary stressed the importance of halting the lake's eutrophication and argued that prompt action in this area could help prevent Lake Michigan from going the way of Lake Erie.[86]

The conference dragged on for nine days. The sheer scope of the testimony and discussion illustrates both the growing scientific knowledge about threats to lake water quality and the widespread desire to control all sources of lake degradation. Thus, the twenty-six formal recommendations that emerged at the conclusion of the conference dealt with relatively new issues such as pesticide contamination, thermal pollution, and watercraft discharges, as well as the old standbys such as combined sewers, the dumping of dredgings, and oil pollution. Following Secretary Udall's lead, the conferees gave top priority to the eutrophication problem.[87]

The subject of treatment requirements and abatement timetables for municipal and industrial dischargers generated the most discussion among the conferees. The heated debate accompanying this issue highlighted the continuing tension between the desire of the federal government to impose standard requirements and the desire of state officials and industry to maintain local control. The FWPCA report prepared for the conference recommended that all municipalities in the basin take action to provide at least 90 percent removal of BOD and 80 percent removal of phosphorus by July 1972. Industries in the basin discharging organic waste and sewage were to achieve the same degree of removal. The state authorities were to determine the waste treatment needs for each industry in the basin within six months and the construction of necessary treatment facilities was to be completed within thirty-six months.[88]

The recommendation for uniform 90 percent BOD removal alarmed the paper producers in the basin, who discharged large amounts of organic waste to the lake and its tributaries. Speaking on behalf of the Wisconsin paper industry, Richard Billings of Kimberly-Clark objected that the recommended treatment requirements amounted to de facto effluent standards, which violated the intention of Congress and disregarded the unique local circumstances of individual plants. Billings also attacked the proposed abatement timetable for industry as being completely unrealistic. Like other business representatives in the region, Billings argued that each of the four states should be given the opportunity to implement the abatement programs accompanying their water quality standards before the federal government imposed any new directives.[89] In the following weeks, letters from some of Wisconsin's major paper producers urged Freeman Holmer, administrator of the Wisconsin pollution control program, to oppose the federal recommendation for BOD removal when the conference reconvened in executive session. The paper firms echoed Billings's arguments, stressing also the impossibility of meeting this standard in many old plants and the excessive costs involved.[90]

Holmer and members of the Wisconsin Natural Resources Board proved receptive to these arguments. They agreed that the federal recommendations for uniform effluent reduction were arbitrary and failed to take into account the conditions of receiving waters, thus violating the philosophy of the Water Quality Act of 1965 and the rights of states to vary the quality of their interstate waters, based on primary uses. As a result of objections from Wisconsin and the other state conferees, the conference's final recommendations dropped the 90 percent BOD reduction requirement and instead stated that all municipalities and industries should be in compliance with their state's interstate water quality standards by the end of 1972. Each state agency was to prepare a detailed abatement plan for each discharger within six months for presentation to the other conferees. The conferees agreed to retain the recommendation for 80 percent phosphorus removal.[91]

The Enforcement Conferences: Promise and Performance

At the close of each session of the Lake Erie and Calumet area enforcement conferences, Murray Stein usually concluded with remarks expressing pleasure at the cooperation of state and local officials and industry. Stein described these areas as the most complicated cleanup projects in the nation, but he emphasized the steady progress being made by the local dischargers and sometimes used the metaphor of seeing the light at the end of the tunnel. In spite of the efforts of Stein, Loring Oeming, and other regulators, however, the water quality in the conference areas showed little signs of improvement and by some indicators actually declined.

In a report presented to the Calumet region conferees in December 1968, Robert Bowden of the FWPCA attributed the lack of water quality improvement in the area to several factors. First, many of the major dischargers had not yet completed their treatment facilities. Second, increased production levels in some cases negated any improvement that might have been expected from a higher degree of waste removal. Finally, combined sewer overflows continued to be a problem, especially since many industries in the area had begun discharging at least part of their waste volume to nearby municipal treatment facilities. The same factors were at work in the shoreline areas of Lake Erie, and failure to reduce phosphorus inputs prevented any progress on the lake's eutrophication problem.[92]

Serious pollution problems remained highly visible in the recreation areas where most members of the public came into direct contact with lake waters. Conditions were particularly bad in the areas that received the highest visitation. In metropolitan Cleveland, for example, a survey conducted by the FWPCA during the 1967 swimming season found that public beaches were "grossly polluted," with adjoining waters registering high bacterial counts and large clusters of algae rotting on the shore. At heavily visited Niagara Falls, a study from the same period found oil slicks on the Niagara River, which was sometimes vividly discolored by industrial effluent. At Niagara Falls itself, the mixing of sewage and industrial wastes produced offensive odors that forced tourists to cover their noses.[93]

Both clean water advocates and local newspapers expressed impatience with the lack of concrete results from the enforcement conferences. In a statement submitted to the June 1968 session of the Lake Erie enforcement conference, congressman Charles Vanik of Cleveland described his frustration with the continuing "talkathon" on the lake's water pollution and noted that water quality was continuing to worsen almost three years after the first session of the conference was convened. Vanik said that the people he represented would not tolerate a "token effort" toward cleanup and promised that "the public protest will not tire or terminate." When the Chicago Department of Water and Sewers reported to the Calumet area conferees in September 1967 that the city's raw water supply from Lake Michigan was continuing to decline in water quality, the headline "LAKE POLLUTION GROWING" stretched across the entire front page of the next day's edition of the *Chicago Tribune*.[94]

Aside from their coverage of the enforcement conference sessions, the major urban newspapers in the Great Lakes region continued to devote feature space to the general water quality problems of the region. In April–May 1966, the *Milwaukee Journal* devoted three consecutive issues of its Sunday magazine, *Picture Journal,* almost entirely to the theme "Pollution: The Spreading Menace." The twenty-seven-page series, composed almost exclusively of graphic color photos detailing polluted conditions throughout the state, won a Pulitzer Prize for the paper. The *Tribune* conducted a Save Our Lake campaign consisting of many articles—often on the front page—detailing pollution problems throughout the Great Lakes. In September 1967, the newspaper collected these articles into a forty-two-page booklet, complete with color photos, which the *Tribune* made available free of charge to interested public officials and citizens.[95]

Traditionally, the government officials responsible for regulating water pollution took a flexible view of abatement timetables. As long as the industry or municipality appeared to be making a good effort, excuses about labor problems, equipment shortages, or technical difficulties usually sufficed. This was especially true in the case of industry. Government officials at both the state and the federal levels perceived recourse to legal action as a violation of cooperative pragmatism, the guiding philosophy of their regulatory efforts. In a May 1966 interview with the *Cleveland Press,* Commissioner Quigley stated frankly that the FWPCA had no plans to take violators to the next stage of the federal enforcement process—the formal hearing board—when deadlines were missed or other requirements were not met. He stressed the role of compromise and negotiation in the enforcement process.[96]

While government regulators and industry took a flexible approach to abatement deadlines, citizen groups, the media, and some politicians viewed them as important milestones in the fight against pollution, which had to be met at all costs. The symbolic importance of these deadlines was especially evident in the Calumet region, where the enforcement conference initially singled out a handful of large industrial plants as the major contributors to the degradation of southern Lake Michigan. The Chicago daily newspapers, which sometimes portrayed the city as a victim of irresponsible out-of-state industries, paid close attention to the progress of these plants. In March 1966, a FWPCA official informed Stein that

> I was positively and firmly informed during my recent Chicago visit that a violation of any of the conference deadlines by as much as 30 seconds would result in a violent and uniform attack on the FWPCA by the Chicago press. Unless our engineers are willing to express their judgement that extensions are necessary, my comment would be one of complete disapproval of any proposed extension.[97]

The issue of deadline extensions for the Calumet area industrial dischargers first surfaced at the March 1967 meeting of the conference. Blucher Poole of Indiana informed the other conferees that the Indiana SPCB planned to approve preliminary construction plans for Inland Steel, Youngstown Sheet and Tube, and U.S. Steel that extended completion into 1970, well past the December 31, 1968, deadline set by the conference for all industrial dischargers. Clarence Klassen of Illinois and Wally Poston, the federal conferee, opposed granting any deadline extensions at this time. Poole maintained that December 1968 had been unrealistic from the first for the huge operations of these steel firms. Stein smoothed things over as usual and decided to put off any decisions on this issue until later in the year.[98]

When the conferees appeared ready in September to approve the steel firms' requests for extension, Mayor Daley sent a telegram to Secretary Udall urging him not to extend the present enforcement conference deadline. The mayor believed the steel mills could meet the timetable with a greater effort: "Instead of them working 8 hours a day, let them work 18 to 20 hours a day as the city does on vital public works." Seven days later, Udall informed Daley in a telegram that he would not extend the enforcement conference deadline. Udall was said to be "highly annoyed" by the steel firms' requests for extensions. Interior staff members pointed out that the Indiana firms had already received two extensions of the deadline for submitting their engineering plans and that the volume of waste emanating from their mills had grown steadily in the last two years.[99]

Secretary Udall's decision may have won praise from Mayor Daley and the Chicago media, but it had little practical effect. Poole continued to maintain that the December 31, 1968, deadline was not feasible. Poston warned that the federal government might take the firms to court if they did not meet the original timetable, but the requirements of the federal law made his threats hollow. Under the enforcement procedures mandated by the Federal Water Pollution Control Act, Secretary Udall would have to convene a formal hearing board and then wait a minimum of six months before the cases could be referred to the attorney general. Given the expected delays of such a procedure, granting the eighteenth-month extension could be expected to bring faster results.[100]

The FWPCA's inability to take prompt action frustrated city of Chicago officials, who felt put upon by the Indiana dischargers. As MSD superintendent Vinton Bacon explained, "Illinois does not gain by eliminating its pollution if the state next to us can continue to pour filth in the lake." At a press conference in October 1967, Bacon announced that the MSD was filing suits against twelve Indiana industries sited on the Indiana Harbor Ship Canal. The district asked the Cook County circuit court to grant a permanent injunction ordering the industries to stop discharges that polluted Lake Michigan. Said Bacon: "The day of chatter is over, the day of using the court is here."[101]

Several days later, MSD attorney Allen Lavin asked Illinois attorney general William Clark to bring suit against the state of Indiana in the Supreme Court for failing to use its police powers to prevent the harmful pollution of Illinois waters by industries and municipalities in Indiana. Lavin emphasized the dangers to public health posed by these discharges and added that similar suits should be instituted against Michigan and Wisconsin "where similar circumstances are prevailing." Two days later, perhaps to demonstrate its impartiality, the MSD filed suit against the South Works of U.S. Steel in circuit court, requesting an injunction that would force the company to cease dumping pollutants into Lake Michigan and the Calumet River.[102] Appearing at a conference on industrial water pollution the following spring, Lavin argued that government pollution control personnel had been brainwashed into thinking it was impractical to vigorously enforce industrial water pollution laws. He called for an end to excuses and maintained that, although pollution abatement laws were not perfect, conscientious staff work and a willingness to back up regulations with legal remedies could bring significant results. Lavin noted that major corporations shunned negative publicity and argued that the prospect of formal hearings and lawsuits usually brought action.[103]

The MSD's strategy was to persuade the industries to settle out of court and agree to a December 31, 1968, deadline for installing treatment facilities, but the court proceedings dragged on through the end of 1968. Illinois Attorney General Clark was hesitant about challenging Indiana in Supreme Court, especially since federal officials pointed out that Illinois was also guilty of polluting Indiana's waters at certain times. Instead, he organized a November meeting of the state attorneys general from the Lake Michigan border states in order to coordinate action. Little came of this meeting.[104]

Concerned citizen groups in the Calumet region continued to oppose any extensions of the December 1968 abatement deadline. A representative from the Save the Dunes council of northern Indiana pointed out that Bethlehem Steel had designed and built a completely new steel complex nearby in less time than these other steel firms claimed they needed to construct and upgrade waste treatment facilities. Murray Stein, however, felt that public condemnation of the large steel mills was misguided in most cases. Stein believed that close examination of the abatement program usually revealed that the company in question had been making a good faith effort and achieving some progress.[105] When the Calumet region enforcement conference reconvened in December 1968 just a few weeks before the deadline for compliance, Stein and the other conferees agreed that teams with representatives from each of the conferees would perform inspections of each plant requesting a deadline extension. The conference would then reconvene in January and come to a decision on each request.[106] At the January session, the conferees agreed to extensions for all of the industries requesting a delay, except U.S. Steel South Works and Republic Steel. The latter discharged into the Calumet River, which did not drain into Lake Michigan. The conferees agreed to turn these two cases over to the Interior secretary for further enforcement action. If things went as planned, the other industries would have their needed treatment facilities in place and running by June 30, 1970, at the latest.[107]

Stein called the state conferees' decision to turn over industrial polluters to Secretary Udall for enforcement action a landmark in pollution control. But it was the legal action of the MSD that was most significant in the long run. Although inconclusive in this case, the MSD's efforts to attain faster and more effective abatement through litigation reflected the sense of frustration felt by many citizens and public officials over the slow pace of cleanup in the Great Lakes region. As the 1960s drew to a close, the public's dissatisfaction with the state of the environment increasingly found an outlet in criticism of industry. A Lou Harris poll, published in 1966, found that the majority of Americans believed most of the local waterways in their vicinity were polluted and that local industry bore the primary responsibility. Clean water advocates in the Great Lakes region blasted industry for its selfishness in despoiling public waters. Even as unlikely a figure as the mayor of Gary, Indiana, joined in the condemnation of corporate greed. At the April 1967 Indiana SPCB hearing on water quality standards for the Calumet area, Gary mayor Martin Katz described the appalling conditions at the city's beaches and charged "that industry has refused to concern itself with . . . the welfare of the people. Industry concerns itself, in too many cases, only with the balance sheet."[108]

Clean water advocates also blamed regulatory agencies for failing to push industry and municipalities hard enough. Many activists had looked to the federal government for more effective regulation, but by the end of 1968 the lack of concrete results stemming from the enforcement conferences had shaken their confidence in Washington. Stein's periodic comments about seeing the light at the end of the tunnel led one newspaper reporter to compare him to an American general in Vietnam. In an article published at the end of 1968, law professor and clean water advocate Arnold Reitze reviewed the efforts of state and federal officials to clean up Lake Erie and concluded that legal action through the courts would have to play a more important role in the future in securing abatement. He noted that relatively few lawyers were involved in shaping and implementing environmental laws and regulations. In light of the record compiled by the sanitary engineers and health officials currently responsible for pollution regulation, Reitze implied that it would be wise to consider a much broader role in the process for the legal profession.[109]

As the following years would demonstrate, many others shared Reitze's sentiments. Cooperative pragmatism, with its reliance on informal negotiation and a balancing of user interests, was seriously out of step with the views of clean water advocates and many influential figures in government and politics, who called for maximum feasible treatment and strict enforcement of abatement deadlines and water quality standards. In addition, waste dischargers were now facing much greater demands than in the past. Critics of cooperative pragmatism argued that informal conferences and moral suasion could go only so far in calling forth the desired expenditures, at least at a rate that was in the public interest. Government regulators and industry officials may have downplayed missed deadlines and continued poor water quality, but clean water advocates viewed these conditions as symptoms of a failed regulatory system badly in need of reform. In the next few years, these reforms would come.

Fire fighters work to put out a blaze on the Cuyahoga River, November 1952.
Courtesy of Cleveland State University, *The Cleveland Press* Collection

Loring Oeming, Executive
Secretary of the Michigan
Water Resources Commis-
sion. Courtesy of Michigan
State Archives

Ohio Governor James Rhodes, left, confers with New York Governor Nelson Rockefeller, far
right, and Kenneth H. Sharpe of the Ontario Water Resources Commission at the Governors
Conference on Great Lakes Water Pollution held in Cleveland, May 10, 1965. Courtesy of
Cleveland State University, *The Cleveland Press* Collection

Lake Erie is dying

Does anybody care?

Picture Lake Erie as 10,000 square miles of lifeless water. A modern Dead Sea. It will happen. How soon? In 20 years? Sooner?

Does anybody care? Eventually all lakes die, but ours is old and feeble before its time. Because we are polluting it to death.

Does anybody doubt this? Walk along its foul beaches and smell the odor of decay. Already game fish—whitefish and walleyed pike—are disappearing from Lake Erie. But scavenger carp hover at the sewer openings, gobbling garbage and filth. Fishermen cannot fish. Swimmers cannot swim. Water sports enthusiasts seek cleaner waters. This is Lake Erie, most polluted of the Great Lakes. A dying lake. Soon it will be too late to save its life. But there is still time. An aroused public opinion can save Lake Erie.

If you care what happens to this area's greatest natural resource, fill out the coupon below and send it to David Blaushild. He'll see that the governor, the Congress and others in authority know that you care.

> Dear Mr. Blaushild. Yes. I share your deep concern about pollution of Lake Erie. I believe we need immediate action to save Lake Erie from destruction.
> name _____
> address _____
> city _____

A public service message from

David Blaushild Chevrolet

16005 Chagrin Blvd, Shaker Heights   Reprint of this ad available on request.

David Blaushild placed this and other advertisements in Cleveland newspapers in 1964.
Courtesy of Ohio Historical Society

Wyandotte Chemicals Corporation discharges industrial waste into the Detroit River, 1966.
Courtesy of National Archives (Record Group 412–series G)

Secretary of Interior Stewart Udall, left, with President Lyndon B. Johnson.
Yoichi R. Okamoto photograph, courtesy of LBJ Library Collection

Murray Stein, standing, and state officials at the June 1968 meeting of the Lake Erie federal enforcement conference. Courtesy of Cleveland State University, *The Cleveland Press* Collection

Protestors outside the June 1968 meeting of the Lake Erie federal enforcement conference. Courtesy of Cleveland State University, *The Cleveland Press* Collection

Illinois Attorney General William Scott. Courtesy of the Illinois State Historical Library

William Ruckelshaus, the first administrator of the federal Environmental Protection Agency, with President Richard Nixon. Courtesy of the National Archives

Sign posted at a Cleveland public beach, summer 1971.
Courtesy of Cleveland State University,
The Cleveland Press Collection

Aerial view of the massive Detroit Wastewater Treatment Plant taken in the early 1970s.
Courtesy of the City of Detroit, Water and Sewerage Department

Part 3

The Breakdown of Cooperation, 1969–1972

CHAPTER FOUR

The New Regulation

Organizational Change and Legal Confrontation

I n the Great Lakes Basin, during the four-year period from 1969 through 1972 the regulatory system of cooperative pragmatism collapsed like a house of cards. Large waste dischargers increasingly found themselves on the receiving end of high-profile lawsuits, and a new generation of regulatory officials emerged to challenge the accepted norms of pollution control and push for faster abatement progress and higher levels of treatment. A number of factors contributed to the breakdown of the established regime's pattern of informal cooperation and voluntarism. First, government at both state and federal levels responded to the increasing complexity of environmental regulation and the public demands for stronger enforcement action by creating new regulatory institutions designed to increase government effectiveness in this area. Just as important, lawyers such as Illinois attorney general William Scott and the first head of the federal EPA, William Ruckelshaus, took on a greater role in pollution control at the federal level and in a number of states. State and federal attorneys showed a new willingness to act independently as enforcers of antipollution laws, while young men with legal backgrounds and orientations assumed the top positions at the new EPA and at some of the state agencies.

The growing number of environmental groups active in the Great Lakes region also contributed to the breakdown of cooperative pragmatism. The seeming inability of government authorities and industrial and municipal polluters to achieve any improvement in Great Lakes water quality allowed environmentalists and their allies to portray this failure as a predictable result of timid government enforcement and industry "capture" of the regulatory apparatus. The shift in regulatory style to public confrontation and litigation was in many respects an attempt to appease these critics, as well as a response to genuine industry and municipal foot-dragging. The highly publicized lawsuits and other enforcement actions of these years partially satisfied environmentalist critics of government water pollution regulation and helped move recalcitrant polluters along the path to significant abatement. Government officials, however, lamented the drain on manpower and other resources that this approach entailed. In addition, the various government agencies at the local, state, and national levels sometimes failed to coordinate their enforcement efforts and related activities, creating an atmosphere of confusion and uncertainty for waste dischargers. The reliance on individual negotiation and settlements following enforcement actions also clashed with the federal government's goal of standardized and equitable treatment requirements.

The Refuse Act permit program was an attempt to restore stability to water pollution regulation, but an unfavorable court decision cut the legs out from under this program before it could get started. State officials bitterly resented the concept of a national permit system. The 1972 amendments to the Federal Water Pollution Control Act, however, established a permit program that was even more ambitious than the system established under the Refuse Act. The 1972 amendments completely revamped water pollution regulation in the United States, granting the federal government primary authority for controlling pollution discharges and creating a system of national discharge permits.

State-Level Reorganization

The reorganization of state environmental programs in the Great Lakes region was part of a nationwide trend. Between 1967 and 1974, thirty-two states—including all of the Great Lakes states except Indiana—consolidated their environmental protection programs within a single department or agency.[1] Several factors were behind this trend. First, the growing realization that decisions made about one form of environmental pollution often affected other problem areas (for example, a greater level of sewage treatment could produce more solid waste). Second, the growing complexity and scope of environmental protection and management, driven in part by new federal mandates and grants-in-aid. And third and most important, the belief among environmental advocates that consolidation of environmental protection programs in a single agency would result in more effective regulation of activities affecting the environment.[2]

The demands for a more active state government in the area of environmental protection coincided with an ongoing historical trend—the increase in power and effectiveness of state governors and their administrations in the decades following World War II. Some of the factors responsible for this transformation—voting district reapportionment, increased two-party competition, and a larger state government revenue base—also worked in favor of greater environmental regulation at the state level.[3] As governors, state legislators, and other elected officeholders and candidates competed for votes in cities and suburban areas, they turned more and more to environmental protection as an issue with popular appeal.

Polling data from the second half of the 1960s indicates the growing importance of environmental issues to voters. For example, in a nationwide poll Opinion Research Corporation found that the percentage of respondents who believed water pollution was "very or somewhat serious" in their vicinity increased
from 35 to 74 percent between 1965 and 1970. Along the same lines, in May
1965 a Gallup poll gave Americans a list of ten domestic problems and asked
them which three they would like to see the government devote most of its attention to in the next few years. "Reducing pollution of air and water" was chosen as
one of the three problems by only 17 percent of the respondents, ranking it
ninth on the list. Presented with the same list in May 1970, 53 percent chose reducing pollution as one of the three most important problems, making it second
only to "reducing the amount of crime" as an object of concern.[4]

Just as important, a number of governors in the Great Lakes states shared a
deep and genuine concern for protecting the environment and gave the issue an
even higher profile through speeches, public appearances, and other activities.
While Democrats in Congress from the Great Lakes region led the calls for a
greater federal role in the regulation of water pollution, moderate or liberal Republican governors in these states were responsible for some of the most innovative water quality initiatives at the state level. Governor Nelson Rockefeller of
New York, who pioneered extensive state funding for sewage treatment plant
construction, even went so far as to publish a book, *Our Environment Can Be
Saved*, in 1970. His younger counterpart, Republican William Milliken of
Michigan, established a national reputation with his environmental initiatives.
Addressing an environmental teach-in on Earth Day 1970, Milliken fully embraced the rhetoric of the environmental movement, which he described as a "social revolution" that would bring "a complete turn-around in the current direction of American life." According to Milliken, "the environmental crisis" was "far
more threatening than any we face."[5]

Across the country, most of the state environmental reorganizations resulted in
either a new agency devoted strictly to environmental protection or a "superdepartment" that combined pollution control with natural resource management
programs. Two states on the Great Lakes set the pattern for reorganization in
1967, when Minnesota created the State Pollution Control Agency and Wisconsin established the State Department of Natural Resources (DNR). Only a handful of states decided to consolidate pollution control programs within a reorganized health department. In the Great Lakes region, Minnesota, Illinois, and
Ohio created environmental protection agencies, while Wisconsin, New York,
Pennsylvania, and Michigan created superdepartments.[6] The environmental reorganizations were sometimes part of a broader restructuring of the state executive
branch. This was the case in half of the environmental reorganizations nationwide. But Wisconsin was the only one of the Great Lakes states in which the reorganization was not environmentally specific, and even here the state's water pollution control apparatus had been completely revamped to improve its effectiveness
the year before the creation of the DNR.[7]

Clean water advocates in the Great Lakes Basin had long been critical of the state agencies responsible for controlling pollution. They pointed to the presence of special interests on the regulatory boards as evidence of the too cozy relationship between regulators and polluters. By the late 1960s, the concept of "regulatory capture" had gained dominance in academic circles, and the general idea that government regulatory agencies were "selling out" the public and catering to industry and other special interests spread to the general public. This idea played an important role in the reforms of the Public Interest Era.[8] State officials like Ralph Purdy, who took over from Loring Oeming as MWRC executive secretary in 1968, continued to extol the benefits of having citizen members on the commission who represented affected interests, but this view was becoming increasingly unfashionable. When the federal EPA started operation in December 1970, administrator William Ruckelshaus began publicly pressuring the states to eliminate from their regulatory boards those members representing polluting interests.[9]

The new state agencies featured full-time directors, usually appointed by the governor, and a much larger staff than the previous single-program boards had enjoyed. Even so, each of the Great Lakes states continued to utilize a citizen board. The tradition of such boards was deeply ingrained in many of the states' political culture. Moreover, some state legislators and other government officials believed that some kind of independent check was needed on executive power, given the importance of the decisions made by these agencies.

The character of the new boards varied widely by state. In Illinois and Ohio, the reorganization created a Pollution Control Board with the authority to review the decisions of the state EPA and set standards and other policies. In both cases, however, membership on the board was full-time, with ample salaries. In other reorganizations, the new boards were part-time and made up either exclusively of citizens (sometimes appointed to represent a particular interest, such as industry), or of both citizens and representatives of state agencies. The authority of each board varied by statute, but the previous tendency of the old single-program boards to defer to the judgment of the administrator was even greater in the consolidated boards, given the range of issues involved and the greater capability of the new agencies.[10]

Some business leaders in the Great Lakes states believed that consolidating responsibility for environmental regulation within one body would lessen confusion and simplify compliance. But the basic rationale for creating the new agencies was to strengthen regulatory control, and the state legislatures usually endowed the new agencies with greater authority and powers than the regulatory boards they replaced. Thus, in some of the states, industry groups formally opposed the creation of new agencies, even when they had participated in the studies preceding the reorganizations. In Ohio, for example, the Ohio Manufacturers' Association was represented on a special Citizens Task Force on Environmental Protection, created in 1971 by newly elected governor John Gilligan to evaluate the state's current environmental protection policies. The task force's final report called for the creation of an environmental superdepartment with no part-time policy board. In a dissenting note to the report, the

Manufacturers' Association argued against "the creation of a large, costly, and unwieldy government agency and program with no clearly stated safeguards against arbitrary and discriminatory action by that agency against individuals or companies."[11]

Under the reorganizations, the sanitary engineers who had administered the state water pollution control programs moved from the health department to the new agencies, along with other staff personnel. But except in the case of Illinois, the engineer-secretaries did not fill the role of director in the new bodies, even though some had experience in air pollution control as well as water quality. One of the reasons given for consolidating state environmental programs under a single administrator was that such a director could serve as a high-profile advocate for environmental protection and provide strong and decisive leadership for the new agency. Circumstances varied across the region, but in several states the new administrators performed such a role.

In Minnesota, the first director of the Pollution Control Agency was a city engineer from South St. Paul whom some groups criticized for not being aggressive enough. But in 1971 incoming Democratic governor Wendell Anderson appointed Grant Merritt to the post. Merritt—a young lawyer formerly active in the Minnesota Environmental Control Citizens Association, one of the state's most active environmental groups—had gained prominence for his leadership of the opposition to Reserve Mining Company's discharge of taconite tailings into Lake Superior. Merritt's appointment sent shock waves through the Minnesota business community, which viewed the young attorney as an uncompromising zealot. And indeed, as director of the Pollution Control Agency, Merritt became a strong public advocate of tough antipollution measures, making frequent public appearances around the state and lobbying the legislature on an array of environmental issues. Merritt also secured the appointment of other environmentalists to positions in the agency and on its governing board.[12]

David Currie, the first chairman of the Illinois Pollution Control Board, shared a similar background with Merritt. After earning his law degree from Harvard University in 1960, Currie served briefly as a law clerk for Supreme Court Justice Felix Frankfurter before joining the faculty at the University of Chicago. Currie authored a number of articles on environmental law and was also an active member of the Clean Air Coordinating Committee, a Cook County environmental group. In 1969, newly elected Republican governor Richard Ogilvie appointed Currie to the Illinois Air Pollution Control Board. Currie became Ogilvie's key advisor on environmental issues and was the primary author of the 1970 legislation reorganizing Illinois's environmental programs.[13] The new law granted the Pollution Control Board responsibility for setting standards and regulations and adjudicating enforcement proceedings, while the Illinois EPA was to serve as the enforcement arm of the state. But in practice the board became the most active and visible of the two bodies in the period immediately following the reorganization, in part because of the strong personalities of the appointed members. Currie in particular was a vigorous proponent of strong antipollution regulation, although he ceased making public statements when business leaders complained that his public posturing clashed with the adjudicative functions of the board.[14]

Henry Diamond, the first commissioner of the New York Department of Environmental Conservation, was a lawyer who had long been active in traditional conservation fields, serving at one point as counsel to Laurance Rockefeller on environmental issues. Diamond became a forceful, highly visible advocate for environmental protection in New York. Like Merritt and Currie, he took a broad view of environmental problems, warning that real progress was possible only if the public was "willing to pay the costs" and "make the sacrifices in terms of horsepower, convenience and number of children." Diamond was thirty-seven years old at the time of his appointment, Currie thirty-four, and Merritt in his late thirties.[15]

In the one instance where a state sanitary engineer became director of one of the new agencies, he was unable to make the transition. Governor Ogilvie named Clarence Klassen as the first director of the Illinois EPA, but in early 1971 the governor asked him to retire. According to two scholars who made a close study of the Illinois reorganization, Klassen's established approach to pollution control, which stressed informal negotiation and voluntary compliance, was at odds with "the new activist, legally oriented leadership" of the Pollution Control Board. Ogilvie replaced him with William Blaser, an industrial management consultant and close ally of the governor. Despite environmentalist concerns about Blaser's commitment to environmental protection, he was an effective administrator.[16]

Federal Reorganization

By all accounts, Richard Nixon had little personal interest in environmental issues. Indeed he deeply distrusted the burgeoning environmental movement, which he lumped together with the antiwar movement and other "anti-establishment" activities. It was ironic, then, that the Nixon administration presided over an unprecedented expansion of the federal government's authority and activity in the field of environmental protection. In large part, the credit belongs to key members of Congress who authored and pushed through to passage landmark legislation such as the National Environmental Policy Act. Nixon, however, did give this area priority, mainly because he and his advisors realized the growing importance of the environment as a political issue. The administration also benefited from the presence on the White House staff of Russell Train, John Whitaker, and other individuals with a conservation background and a sincere interest in environmental protection. Key presidential aide John Ehrlichmann used his influence to shield environmental initiatives that came under criticism from other quarters.[17]

Still, Nixon's environmental program got off to a rough start when his nominee for Interior secretary, former Alaska governor Walter Hickel, made some intemperate comments criticizing what he viewed as misguided government conservation policies. These remarks, and Hickel's close political ties to development interests, opened the Alaskan to a barrage of criticism from environmentalists in and out of Congress and subjected the former governor to a grueling confirmation process. Hickel later surprised his critics with a number of pro-environment policy decisions, but in the field of water pollution control he did not play the prominent role that Stewart Udall had. Instead, Hickel delegated decision-making authority to Carl Klein, assistant secretary for Water Quality and Research.[18]

Before his appointment, Klein served in the Illinois legislature. In 1967, the Illinois Izaak Walton League awarded Klein its Legislator-of-the-Year Award for his leadership of the legislative Water Pollution and Water Resources Commission. A native of Chicago, Klein carried his personal interest in preserving Lake Michigan with him to the Department of Interior. Klein, a large bull-necked man with a prominent bald head, possessed a no-nonsense style that some found overbearing. The former Army staff officer quickly established firm control over the FWPCA and its new commissioner, David Dominick. The thirty-two-year-old Dominick was a purely political appointee. He lacked experience in water pollution control but had the right political connections. Klein, fifty-two years old and a product of Chicago's Central YMCA Community College, seemed to enjoy lording it over the Yale-educated Dominick.[19] Klein initially discussed the need for the federal program to rely less on enforcement conferences and strict statutory compliance and more on informal, private negotiations with dischargers. This view was seriously out of step with the views of important members of Congress and influential segments of the population, however, and several months later the assistant secretary was making public pronouncements about the need for the FWPCA to initiate more enforcement actions, including legal suits. Klein subsequently reorganized the administrative structure of the agency in an effort to strengthen its enforcement capability.[20]

The Nixon administration's most significant environmental initiative may have been the creation of the federal EPA. In November 1969, the President's Environmental Message Task Force proposed the creation of a federal superdepartment that would combine pollution control functions with natural resource programs. Nixon was sympathetic to the concept, but the alternative of a new regulatory agency consolidating all of the government's pollution control programs ruffled fewer feathers and still offered the promise of greater efficiency and effectiveness. On July 9, 1970, President Nixon submitted his proposal for the creation of the EPA to Congress. Congress failed to take action to reject the reorganization plan, and the new agency began operation on December 2, 1970.[21]

The man picked to head the EPA, William Ruckelshaus, was a relatively unknown thirty-eight-year-old lawyer in the Justice Department. Ruckelshaus came from a prominent Indianapolis family that had long been active in Republican politics. After a brief stint in the Army, Ruckelshaus graduated from Princeton University and then earned his law degree at Harvard Law School in 1960. After passing the Indiana Bar, the young lawyer gained his first experience in pollution control when he was appointed deputy state attorney general and assigned to the Indiana Board of Health. One of his duties was assisting the Indiana SPCB in enforcement.[22] After four years in the state attorney general's office, Ruckelshaus entered politics, first losing to a more conservative opponent in the primaries for a congressional seat in 1964 and then winning election to the Indiana House of Representatives two years later. Ruckelshaus became Majority Leader in his first term and impressed enough people to be named the Republican nominee for the U.S. Senate race in 1968. He lost to Democratic incumbent Birch Bayh but was then selected for a Justice Department post by new attorney general John Mitchell, whom he had met during the campaign. Ruckelshaus was serving as assistant U.S. attorney for the Civil Division when, on Mitchell's recommendation, Nixon appointed him to head the EPA.[23]

As the first administrator of a new and important federal agency, Ruckelshaus had the opportunity to shape the EPA's mission, organization, and image. To this day, the agency culture at the EPA reflects his personal stamp. When Ruckelshaus assumed his new position, public concern about the environment was peaking, but so was disillusionment with government efforts to protect the environment. This dissatisfaction was part of a broader distrust of American institutions that had been building throughout the 1960s and that became most visible in public opposition to the Vietnam War. Ruckelshaus later recalled that his first priority when he became EPA administrator was to convince people that the new agency was indeed going to respond to public concerns and take strong action to defend the environment, including tough enforcement action against some of the country's largest corporations. Above all, he believed, it was vital to demonstrate to the American people that progress was being made in this area. Ruckelshaus looked to "public opinion"—that is, the environmental movement and the media—for political support during the early days of the agency. This constituency demanded strong and visible enforcement action, and the new administrator was determined to give it to them. A tall, broad-shouldered man with dark hair, horn-rimmed glasses, and a square jaw, Ruckelshaus succeeded in projecting an image of integrity and purpose in his public appearances.[24]

Frustration in the Great Lakes Basin

Ruckelshaus's perception of growing public dissatisfaction with government efforts to clean up the environment was especially accurate in the Great Lakes region. As the 1960s drew to a close, there was little evidence of improvement in the most degraded areas in the basin. And the widespread lack of compliance with the abatement timetables established by the federal enforcement conferences and the state agencies added to the sense of inaction. At the August 1969 meeting of the Calumet area federal enforcement conference, a technical committee appointed by the conferees reported that water quality in southern Lake Michigan and the rest of the conference area continued generally unsatisfactory and would remain so until the major steel plants completed their remedial treatment facilities. Murray Stein was dismayed at the findings but remained optimistic that progress was being made and that significant improvements in water quality would be achieved when the major abatement projects began to come on line.[25]

When the Lake Erie enforcement conference reconvened in June 1969, George Harlow of the FWPCA's Lake Erie Basin office reported that approximately 75 percent of the cities and 47 percent of the industries had fallen behind the abatement schedules established by the conferees. Stein was unhappy with these figures, but he explained that the conferees needed to consider each case individually and decide if a good faith effort was being made by the discharger and whether a timetable extension was reasonable, given the circumstances. Based on the state reports, he believed that most of the dischargers were acting in good faith and that they could be considered in "substantial compliance" with the conference requirements. Stein also argued that if the overriding goal was to clean up Lake Erie, the conferees had to ask themselves if further legal action would really result in faster progress toward this goal.[26]

Many concerned observers lacked Stein's patience, including some vocal members of Congress from the Lake Erie Basin. In a submitted statement, Representative Richard McCarthy of New York emphasized the lack of any concrete progress after four years of meetings and scored the state agencies for their willingness to grant extensions and exemptions to industries faced with enforcement orders and other regulations. McCarthy's congressional colleague from Cleveland, Charles Vanik, submitted his own statement that criticized the conferees for engaging in strong enforcement action only in cases involving relatively minor dischargers. Vanik noted, correctly, that the cases turned over to the state attorneys general for prosecution usually involved truck stops, small "garage shop" manufacturers, and other minor dischargers. Conversely, large, recalcitrant polluters such as Hammermill Paper's Erie plant received repeated extensions while regulatory officials engaged in protracted negotiations with company officials. Real progress would come, Vanik argued, only when the largest sources of pollution were subject to the same pressures as "the little guys."[27]

The comments of these two legislators accurately reflected the views of many of their constituents and other concerned citizens throughout the Great Lakes Basin. Aside from polling data, another quantitative measure of the growing popular interest in the environment was the growth in the membership of existing conservation organizations and the founding of new groups. At the national level, for example, membership in the Sierra Club grew from 35,000 to 147,000 between 1966 and 1975, while the Audubon Society increased its size from 45,000 members to 321,000 during the same period. By the end of the 1960s, the term "environmentalist" had come into common usage, signifying a new awareness of the delicate nature of the ecosystem and human society's capacity to inflict lasting damage on the planet. Many of the new organizations concentrated almost exclusively on pollution control issues, while some established conservation groups— such as the Sierra Club—also devoted much of their activity to these issues.[28]

In the Great Lakes Basin, the number of citizens actively involved in environmental issues also swelled. Much of this growth took place in community groups that formed in response to particular problems of local concern. In the heavily polluted environs of Gary, Indiana, in the 1960s there was tremendous growth in the membership of existing environmental groups, and concerned citizens also established new environmental organizations. The Lake County Fish and Game Association, for example, had been founded in 1920 and served primarily as a social club. But after the association became active in the fight against Lake Michigan water pollution, membership expanded from about 30 in 1961 to 300 in 1969. Within Gary city limits, the Glen Park Izaak Walton League also expanded its membership as the group became more active in antipollution lobbying efforts.[29] New groups experienced even faster growth in membership. In February 1969, a group of citizens in northeastern Minnesota formed the Save Lake Superior Association after Interior Secretary Udall announced the convening of a federal enforcement conference on Lake Superior water pollution. By the time the conference opened in May, the association had grown to 800 members in the three states bordering the lake.[30]

A young female college student initiated the formation of the Save Lake Superior Association. This was not unusual. Across the Great Lakes Basin, women played a leading role in the organizations working to promote restoration of Great Lakes water quality. In Gary, the local chapter of the League of Women Voters became heavily involved in community pollution issues, as did many league chapters across the region. In 1969, some members of the Gary chapter decided to form a new group, Community Action to Reverse Pollution (CARP), that would devote itself primarily to environmental issues. According to a close observer of Gary environmental issues, CARP became the most active environmental organization in the Gary area. While men composed almost half of CARP's membership by 1972, women continued to hold most of the leadership positions in the group. In the Buffalo area, Housewives to End Pollution consisted of women "who have banded together to attack immediate local pollution problems that center around the home."[31]

These and other groups that sprang up across the shores of the Great Lakes—Citizens for Clean Air and Water (Cleveland), the Lake St. Clair Anti-Pollution League (Michigan), the Citizen's Committee on Pollution (Buffalo), and many others—engaged in traditional lobbying and educational activities. They testified at the enforcement conferences and other public hearings. They sent letters and petitions to elected officeholders and invited government officials to appear at their meetings. They took pictures and collected water samples to turn over to regulatory officials and embarrass dischargers.

But in spite of these efforts and the increased activity of government officials, there seemed to be little improvement in area water quality. The existing cynicism about government regulation of pollution became even more pronounced. The increased federal role that had been eagerly sought earlier in the decade had not brought the hoped-for results. By 1970, even a representative from the usually staid League of Women Voters was condemning the "gigantic inefficient [regulatory] bureaucracy" that had only produced "talk and more talk."[32] Many environmentalists in the region believed that regulatory officials were too quick to accept the explanations offered for delay by corporate dischargers lagging behind their abatement schedules. These critics argued that in many cases the firms in question had simply made an economic decision to stall for as long as possible before making needed capital expenditures. Environmentalists maintained that only aggressive legal action, including criminal prosecution of corporate officials and temporary plant shutdowns, could force the most recalcitrant polluters to act in the public interest.[33]

The increased cynicism of concerned citizens in the Great Lakes Basin and their desire to make both government and business more accountable to the public interest were part of a larger trend in American society. The diverse social movements of the 1960s—civil rights, protest against the Vietnam War, consumer rights, feminism—all had in common demands for increased accountability and participation, which they directed at the major institutions of American life.[34] Large business corporations were a primary target of criticism, as their public image declined sharply in the late 1960s and early 1970s. A Yankelovich survey found that the proportion of the public who believed "business strikes a fair balance between profits and public interest" dropped from 70 to 20 percent between 1968 and 1974. To many citizen activists, government regulatory agencies appeared to be more interested in protecting corporate interests rather than the public interest. The emergence of public interest law was one attempt to redress this imbalance.[35]

Public interest law and the activities of self-described public interest organizations had a profound effect on the environmental movement. David Vogel has distinguished the public interest movement from the mainstream of the "New Deal left-liberal tradition" by the former's lack of confidence in either public or private bureaucracies. Those active in the public interest movement sought to decentralize the regulatory system and make both the corporate and the government sectors more accountable to the public. Expanding the opportunities for public participation in the activities of both these sectors offered one avenue for achieving these objectives. Through court action, citizens could exert a direct influence on government and business policy, and environmentalists sought to do just that with the establishment of environmental law firms.[36]

In 1967, a group of lawyers and scientists from Long Island, frustrated by the limitations of public education and lobbying, formed the Environmental Defense Fund. This initially small organization combined leading-edge scientific research with innovative legal tactics in an effort to bring about concrete changes in public policy toward the environment. The Environmental Defense Fund achieved a number of early successes, including an eventual ban on the use of DDT. The National Resources Defense Council, founded in 1970, and other environmental law groups followed in the path of the Environmental Defense Fund and also made their mark on environmental policy in a number of important areas. The influence of these legal organizations on the traditional conservation groups was just as important as their direct achievements. In the Great Lakes states, environmental groups made the initiation of or active involvement in environmental litigation an important aspect of their mission. This development also reflected the growing use of litigation as an enforcement tool by state and federal officials.[37]

One organization in the Great Lakes region that came to play an important role in the region's legal conflicts over environmental issues was the Chicago-based Businessmen for the Public Interest (BPI). This organization, established in 1969, was a nonprofit corporation made up of a small group of young lawyers, research associates, and a board of directors from the forty Chicago business concerns that provided part of the organization's operating budget. These directors were primarily small businesspersons: retailers, bankers, small manufacturers, investment brokers, and the like. Prominent Chicago civil rights lawyer Alexander Polikoff served as BPI's executive director. Aside from improving the environment, BPI included among its objectives liberal goals such as providing relief for the poor and distressed, lessening neighborhood tensions, and eliminating prejudice and discrimination. BPI relied on legal action to achieve these objectives; its lawyers appeared in court or before regulatory agencies and tried to force compliance with the law or win new legal interpretations that would require corporations or the government to take some kind of action. Although active in the campaign to effect faster abatement of industrial and municipal waste effluent, BPI had its greatest impact on the controversy over thermal pollution from electric utility power plants.[38]

Established conservation groups in the Great Lakes Basin had long been clamoring for greater use of the courts in controlling pollution. In 1965, Wayne Harris, a prominent conservation leader in western New York and a member of the Monroe County Conservation Council (Rochester), proposed that legislation be adopted enabling private citizens to initiate injunctive proceedings through the state attorney general when they could establish proof that coliform readings in state waters exceeded accepted safety levels. That same year, Cleveland-area activist David Blaushild filed a taxpayer's suit against city officials in an effort to gain enforcement of the city's antipollution ordinances. Despite these and other early efforts, citizen involvement in the legal enforcement of antipollution laws remained minimal because of traditional restrictions on legal standing, lack of sympathy from regulatory officials, and a general lack of expertise, or at least boldness, on the part of conservationists.[39]

By the late 1960s, however, many government officials also embraced the concept of increasing citizen participation in the regulatory process. In Wisconsin, which possessed a long tradition of citizen involvement in government, the law creating the Wisconsin DNR in 1967 also provided that legal standing would be granted to six or more citizens who filed a complaint with the DNR regarding alleged or potential pollution. The DNR was then required to hold a formal hearing on the complaint, and the agency's decision could be appealed to the county circuit court. As part of his effort to restore public confidence in government, EPA Administrator Ruckelshaus called on citizens to initiate "responsible" court action against polluters and to bring pressure against all levels of government to enforce existing pollution laws.[40]

At the national level, a series of federal court rulings in the late 1960s and early 1970s gradually broadened the traditional bases for legal standing to sue. At the state level, a number of legislatures passed laws designed to make it easier for citizens to engage in legal action to protect the environment. Michigan pioneered in this area with the passage of the Environmental Protection Act of 1970. This statute—sometimes referred to as the "Sax Law," after its original author, University of Michigan law professor Joseph Sax—was the first law to provide citizens with a statutory right to sue to protect the environment for the general good. The Michigan law inspired similar statutes in Indiana, Minnesota, and other states.[41]

Sax's views represented the thinking of many environmentalists during this period. In his influential book, *Defending the Environment,* published in 1971, Sax argued that members of the public had given up control over the treatment of their natural environment to professional experts, who determined policy in a process closed to the public. Powerful and organized economic interests, on the other hand, had learned the skills to manipulate legislative and administrative agencies. He believed that the courts, because of their status as outsiders in the normal regulatory process, were more impartial and less susceptible to external pressures. In court, the individual citizen or community group could obtain a hearing on equal terms with more powerful interests. Sax called on Americans to assert their legal right to the maintenance of a clean, safe environment.[42]

The Sax Law recognized the increased willingness of environmentalists to use litigation as a tool to advance their environmental agenda. In October 1971, the Izaac Walton League executive board, recognizing that the organization's financial resources were no longer adequate to support all of the league's legal efforts, adopted a resolution creating an Izaac Walton League legal action fund. Since environmental lawsuits had become "a vital expression of League purposes," it was essential to gain increased funding for these activities. Between March 1969 and February 1972, the national league and its state divisions and chapters brought suit twenty-seven times to protect the environment.[43]

From a broad perspective, however, the environmental groups in the Great Lakes region played their most important role in the legal sphere through their specific assistance in government suits and their general demand for such legal action. At one point, Illinois attorney general William Scott, who established a national reputation through his highly publicized legal actions against major polluters, claimed that at least half of the environmental protection cases filed by his office were the result of citizen complaints and that these cases were further assisted by citizen testimony in court. In addition, when environmentalists like Grant Merritt gained positions in state regulatory agencies, they brought their propensity for legal action with them. Joseph Karaganis of BPI served as a consultant to Scott before being named chief of the Illinois attorney general's environmental control division in 1972.[44]

State-Level Enforcement

During the second half of the 1960s, state water pollution control agencies in the Great Lakes region began to make greater use of formal administrative orders against dischargers, including detailed abatement timetables and specific effluent discharge requirements. In many cases, the states were responding to the requirements of the federal enforcement conferences. The commitment to meeting these conference objectives varied by state, with Ohio officials being notable for their willingness to extend final compliance deadlines. State regulators acting outside the jurisdiction of the federal enforcement conferences also experienced widespread noncompliance with abatement schedules. In New York, an inquiry by the *Buffalo Evening News* early in 1971 revealed that more than one out of three dischargers in the Niagara River–Lake Ontario Basin were behind on state-mandated abatement schedules.[45]

Despite these delays, state sanitary engineers continued their ingrained aversion to litigation. They agreed with Clarence Klassen of Illinois that lawsuits should be "an absolutely last resort" in pollution control because of all the delays inherent in the process. And keeping out of court meant keeping control of the case in the administrative agency.[46] Moreover, the engineers and their staffs blamed many of the cleanup delays on the failure of the federal government to follow through on its promises for dramatic increases in construction grant funding. This also affected industry, since more and more plants were making agreements with municipalities for joint waste treatment.

In spite of these misgivings, the state agencies in the Great Lakes Basin were gradually showing a greater willingness to refer cases to the state attorney general for enforcement action. In New York, for example, as of August 1971, twenty-eight of the original forty-four tabulated sources polluting Lake Erie had still not cleaned up their operations. Of that twenty-eight, Department of Environmental Conservation staff had referred six municipalities and six industries for departmental penalty assessments or legal action by the state attorney general. In this instance, however, most of the cases involved small villages and food processors.[47]

Environmentalists were not the only ones who criticized state agencies for their concentration on relatively minor polluters. In February 1970, Wisconsin attorney general Robert Warren appeared before the State Natural Resources Board to criticize the Wisconsin DNR for referring only five water pollution cases to his office in the year since his election. To make matters worse, these cases had all been small municipalities and industries. Warren said it was time for the state to begin going after major polluters with "a missionary zeal." He pledged that his office would act independently, where appropriate, to bring action against polluters under public nuisance statutes.[48] Other state attorneys general in the Great Lakes region expressed similar sentiments. As elected law enforcement officials, they had the incentive and powers to respond to the public clamor for tough enforcement action against polluters. By far the most active state attorney general in the Great Lakes region on this issue was William Scott of Illinois.

Scott, like Warren, was a Republican. According to Scott, when he assumed office early in 1969 and examined past state efforts against polluters, he found that, except in cases involving fish kills or other overt discharge violations, the defendants were generally individuals or relatively small enterprises. Scott was determined to concentrate his efforts on the major sources of pollution. The Chicago-born lawyer decided also that too often the voluntary agreements worked out in informal conferences between government and industry were later violated or deadlines freely extended. To avoid this type of outcome, Scott would proceed against the polluters in court, where promises were binding and the public could keep a watchful eye. In order to make this possible, Scott's office sought and obtained new statutory powers that allowed the attorney general to act independently of other government proceedings in order to prevent air and water pollution. In August 1969, Henry Caldwell, Scott's assistant attorney general, appeared before the Calumet area federal enforcement conference and announced that the time for deadline extensions was over. Referring to the recent legislation, Caldwell pledged that the attorney general's office would use every judicial process at its disposal to abate water pollution.[49]

Earlier in the year, the Calumet area conferees had referred the cases of U.S. Steel South Works and Republic Steel to the Interior secretary for further enforcement action. Interior took no action, however, and when the conference reconvened on August 26, Assistant Secretary Klein began the session with harsh criticism of the conferees for referring the cases to Interior, given the time-consuming and cumbersome enforcement procedures that the federal government was forced to follow. Klein made it clear he preferred to leave it to the Chicago MSD and Illinois to take the appropriate action. The *Chicago Tribune* scored Hickel and Klein for their refusal to take action against the companies. "The Federal government seemingly doesn't want to get involved in water pollution control on the Great Lakes," Caldwell remarked afterward.[50]

On September 18, 1969, Attorney General Scott filed a civil action against U.S. Steel South Works. Several weeks later, the state suit was combined with the suit initiated by the MSD against U.S. Steel in October 1967. When the Illinois Sanitary Water Board held its regular meeting in October, the board decided to deny the requests for further compliance deadline extensions filed by U.S. Steel, Republic Steel, and Interlake Steel. The members then agreed to request Scott to initiate legal action against these three firms, along with the MSD itself and three other plants discharging into the Calumet River system. All of these dischargers had failed to meet the September 30, 1969, abatement deadline established by the board.[51]

Scott's Republican colleague, Illinois governor Richard Ogilvie, was also attracted to the concept of increased litigation to accelerate pollution abatement. Governor Ogilvie thought big. In late October 1969, Ogilvie requested Scott to file suit in the U.S. Supreme Court against Wisconsin, Indiana, and Michigan in an effort to halt the further degradation of Lake Michigan. The governor also wanted to include the major industries and municipalities of those states as defendants in the suit. The Lake Michigan federal enforcement conference had already devised a comprehensive abatement plan for the lake, but Ogilvie said his action was necessary to "stop the empty rhetoric, the foot-dragging, the confusion and the uncoordinated efforts of the past." Perhaps thinking back to the positive results of Illinois's Supreme Court suit against Indiana in the 1940s, Ogilvie anticipated hearings before a court-appointed Special Master and the issuance of a comprehensive order that would set down clear and certain steps for compliance by each discharger. Scott rejected Ogilvie's grand scheme as impractical, but subsequently demonstrated his willingness to go after polluters in neighboring states.[52]

In February 1970, Scott filed suit against twelve municipalities and industries in Illinois and Indiana (including Inland Steel, Youngstown Sheet and Tube, and Standard Oil of Indiana) on pollution violations going back to 1967. Scott brought these actions under common nuisance laws that were unrelated to any government agency timetables.[53] Later in the year, after a meeting of the Lake Michigan federal enforcement conference publicized Milwaukee's failure to implement chlorination of the city's sewage as required, Scott requested leave from the U.S. Supreme Court to file suit against Milwaukee and other Wisconsin cities that were failing to disinfect their effluent. Several Illinois beaches along the state shoreline north of Chicago had been forced to close the previous year because of high coliform levels. Wisconsin state authorities vehemently denied any culpability, citing Milwaukee's efficient secondary treatment that removed over 96 percent of fecal coliform from the sewage prior to discharge, but Illinois officials preferred to take no chances when it came to protecting Chicago-area beaches. Scott wanted the U.S. Supreme Court to hear the case in order to avoid the delays and inevitable appeals that would follow proceedings in the lower courts.[54]

The Supreme Court refused to hear the Milwaukee suit, forcing Scott to pursue the case through a federal district court, but the Illinois attorney general had more success in some of his other actions. After long months of negotiation and testimony, the suit against U.S. Steel was resolved in January 1971 when the company agreed to accept a court-directed timetable for installing recycling equipment at its South Works plant in Chicago. By this time U.S. Steel had met most of the original treatment requirements mandated by the federal enforcement conference, but Scott's agreement required the firm virtually to eliminate the discharge of pollutants into Lake Michigan. The timetable, established by the Cook County Circuit Court, mandated the complete elimination of all ammonia, cyanide, and phenol discharges by October 1972. By November 1975, all other wastes were to be either recycled through the water systems of the South Works or routed into the MSD disposal system. Experts estimated the cost of installing the recycling facilities as somewhere between $8 and $12 million.[55]

Scott called the decision "one of the biggest industrial victories in the country," but Chicago-area environmentalists were not so quick to praise the agreement. The following month, the Izaak Walton League, the Committee on Lake Michigan Pollution, the League of Women Voters, and BPI filed a petition in circuit court requesting the court to nullify the agreement because all the meetings prior to the settlement had been closed to the public. A representative from BPI also criticized vague wording in the decree. The reaction of the environmentalists illustrated both the rising expectations of public interest groups and their distrust of government.[56]

Scott later won a similar commitment from major out-of-state dischargers. Inland Steel completed construction of a $10 million terminal waste treatment plant in 1970, as mandated by the Indiana SPCB and the Calumet area federal enforcement conference. This did not satisfy Scott, who filed civil suit against Inland in January 1972 for polluting Illinois's Lake Michigan waters. The suit came to trial in February 1973. In June, Inland announced plans to implement a new $14 million abatement program that would be completed in 1976 and result in the recycling or deep-well disposal of much of the plant's wastes. Scott continued the suit against Inland, but he was able to complete a circuit court agreement with Youngstown Sheet and Tube that was also based on complete waste recycling.[57]

Company executives found Scott's approach unsettling. Officials from the attorney general's office contacted Abbot Laboratories about reducing waste discharges to Lake Michigan from the company's North Chicago plant, and after a series of meetings with Scott's assistants Abbot executives agreed to a $2.5 million remedial abatement program. When Abbot Laboratories publicly announced plans for the new program, company officials apparently thought that would be the end of it. They received a shock the next day when Attorney General Scott announced at a press conference that he would be filing suit against Abbot for allegedly discharging waste that forced the closing of some Lake Michigan beaches. Scott said he was well aware of the company's proposed improvement program, but he wanted a court-enforced abatement timetable.[58]

Other state attorneys general in the Great Lakes region also responded to the public clamor for stronger enforcement of antipollution laws, although none were as active as Scott. In Ohio, a federal EPA official reviewing the state's water pollution control program commented that the aggressive participation of Ohio attorney general William J. Brown was "the single most important factor in the increase in enforcement related activities." Like his counterparts in other states, Brown created a special division within his office to work full-time on environmental cases.[59] The attorneys general made sure that their activities in this area were well publicized, and not just to place further pressure on polluters. A week before the New York state election of 1970, New York attorney general Louis J. Lefkowitz ran a large campaign advertisement in the *New York Times* extolling his environmental record. The ad claimed that in the mid-1960s Lefkowitz had created in his office a special unit to attack pollution problems—the first in any attorney general's office in the country. It then went on to mention the thirty-nine judgments the attorney general's office had obtained against municipal and commercial polluters. One of these actions resulted in a $10,000 fine against Mobil Oil's refinery on the Buffalo River. A few days before his ad appeared, Lefkowitz announced at a press conference that he had asked the state supreme court to order the refinery shut down to prevent further discharges.[60]

Federal Enforcement

Carl Klein had declined to take further enforcement action against major dischargers in the Calumet area adjoining southern Lake Michigan. But several days after Klein called on state and local authorities to take care of their own problems at the August 26, 1969, meeting of the Calumet area federal enforcement conference, his superior, Interior Secretary Hickel, decided to initiate unilateral federal enforcement action against five major sources of pollution along the Ohio section of the Lake Erie shoreline. On August 30, Secretary Hickel sent letters to the city of Toledo, to Interlake Steel's Toledo plant, and to the Cleveland mills of U.S. Steel, Republic Steel, and Jones and Laughlin Steel notifying them that they were in violation of state and federal water quality standards. The violators had 180 days in which to resolve the problems before facing suit in federal court. This was the first time the federal government had made use of the enforcement provision in the Water Quality Act of 1965.[61]

The 180-day notices, as these actions came to be called, were based on the failure of the dischargers to comply with Ohio's water quality standards implementation plan. Except in the case of Interlake Steel, each violator had missed a deadline for gaining approval of construction plans. Subsequent to the notice, FWPCA investigators uncovered specific pollution problems at each plant that were previously unaddressed by the remedial programs. They made these also part of the violation. Citing the legislative history of the Water Quality Act, federal officials followed up on the notices with informal conferences with each discharger. Except for complaints about the arbitrary nature of the actions and some initial resistance from Republic Steel, representatives of Toledo and the four industries cooperated fully with the FWPCA, and the meetings resulted in a commitment from each discharger to correct the specific problems uncovered by investigators and to proceed with the remedial abatement program as rapidly as possible. The federal government took no further action against the violators at the end of the six-month period. George Harlow of the FWPCA's Lake Erie office expressed his satisfaction with the outcome and hoped that concentrating the enforcement actions in such a high-profile area would help speed up individual abatement programs across the nation.[62]

Ohio state officials, however, expressed outrage at this unilateral federal action. Appearing at one of the informal conferences, health department director Emmet Arnold described the actions as "a waste of time and money and . . . a serious breach of faith with Ohio's water pollution control board's programs." Arnold pointed out that state authorities had approved each of the abatement extensions as reasonable adjustments and were generally satisfied with the progress being made at the plants. Governor James Rhodes backed up his health director, first meeting with Secretary Hickel and then holding a Washington press conference where he criticized federal officials for failing to consult with their counterparts in Ohio prior to issuing the violation notices.[63]

Environmentalists were disappointed at the failure of federal officials to follow up with further legal action on the 180-day notices. Court action under the Water Quality Act was not as simple as it seemed, however. Unless the governor consented to legal action against the dischargers (which appeared very doubtful in this case), the federal government would be required to present firm evidence that the standards violations resulted in harm to the health and welfare of citizens in another state. This would be especially difficult in the cases of the Cleveland dischargers. In the view of environmentalists, complicated procedural requirements had once again prevented quick, effective legal action against polluters.

Some influential figures in the Nixon administration shared the environmentalists' frustration with slow-moving, complicated enforcement procedures. In November 1969, the President's Environmental Message Task Force, led by John Whitaker, presented a preliminary report to President Nixon. Stronger federal enforcement of the nation's water pollution control laws was one of the task force's major recommendations. Nixon embraced the task force's recommendations, and in his February 1970 Message on Environmental Quality, the president proposed a series of amendments designed to streamline and strengthen federal regulatory and enforcement powers in the area of water pollution control. One key proposal provided for individual effluent standards to complement the state-federal water quality standards. In addition, the Interior secretary would be granted review authority over state water quality standards on all navigable waters, not just interstate waterways. Other proposals were designed to make it easier for the Interior secretary to initiate effective enforcement action, including injunctions in emergency situations and fines of up to $10,000 a day for failure to comply with water quality standards or implementation schedules. Despite Nixon's proposals, it would be over two years before Congress amended the basic federal water pollution law, while the relevant committee chairs focused their attention on air pollution control.[64]

Ironically, it was an obscure law from the previous century that finally provided federal officials with a straightforward, rapid means of taking polluters to court. The Refuse Act, as it came to be known, was actually a section of the Rivers and Harbors Act of 1899. This section prohibited the discharge of refuse into navigable waters without a permit from the Army Corps of Engineers, excluding liquid waste from streets and sewers and activities carried out by the U.S. government in constructing public works. Violators of the act were subject to a fine between $500 and $2,500, or a prison term between thirty days and one year, or both a fine and imprisonment. The courts and other authorities initially interpreted the statute as applying only to cases in which navigation was impeded. While later court interpretations held that the law did apply to dumping per se, regardless of the effect on navigation, very narrow readings of what constituted "refuse" and "dumping" prevented widespread application of the law. But during the 1960s, several Supreme Court decisions authored by Justice William O. Douglas transformed the Refuse Act into a potentially powerful tool for protecting water quality. In these cases, the Supreme Court's majority opinions read the act broadly as a prohibition on the dumping of "all foreign substances and pollutants" except municipal sewage. The federal courts subsequently incorporated this interpretation into their rulings on the Refuse Act.[65]

Environmental activists and some government officials were impressed with the simplicity and directness of the Refuse Act. In September 1969, U.S. attorney Thomas A. Foran of Chicago asked the Interior Department for information on three industrial polluters in the Calumet area—U.S. Steel South Works, Republic Steel, and Ford Motor Company—for use in possible Refuse Act suits against these firms. Assistant Secretary Klein instructed FWPCA Commissioner Dominick not to provide the requested data. Both Klein and the Interior Department counsel for water resources believed that federal action under the Refuse Act might interfere with Illinois Attorney General Scott's suits against the firms.[66]

Foran did not give up. The following February, under the Refuse Act, he filed criminal suits against eleven firms in the Chicago area for depositing waste materials into local waterways. The Army Corps of Engineers, which was working to develop a new reputation as an environmentally sensitive agency, referred the cases to the Justice Department. Later that same month, the Justice Department filed criminal suit against U.S. Steel's massive Gary Works for illegally discharging waste material, primarily oil, into the Grand Calumet River and Lake Michigan on October 11, 1967. Again, the Army Corps of Engineers supplied the evidence. U.S. Steel was later convicted in a jury trial and fined $5,000. The following month, a federal grand jury initiated by U.S. Attorney Foran returned a five-count criminal indictment against U.S. Steel South Works and its plant superintendent for discharging blast furnace wastes into Lake Michigan the previous September. U.S. Steel eventually pleaded guilty and paid a fine of $7,500. The court dismissed the charges against the superintendent.[67]

These convictions may have provided environmentalists with some satisfaction and generated negative publicity for the accused firms, but a company like U.S. Steel could pay these fines from its petty-cash fund. Other federal actions under the Refuse Act, however, demonstrated the broader potential of the law. In March 1970, Secretary Hickel brought the Refuse Act to national attention when he used it to sue Florida Power and Light to prevent thermal discharges from its new plant into Biscayne Bay. In this case, Hickel made the action a *civil* suit in an effort to obtain an injunction against thermal discharges from the utility. Over a year later, the company entered into a consent decree with the government that met Hickel's concerns. The Refuse Act did not expressly provide for injunctive relief, but the Supreme Court decisions that had redefined the scope of the act also ruled that the federal government could obtain injunctions under the legislation.[68]

Federal officials made dramatic use of the Refuse Act in the summer of 1970 when revelations about the potential dangers from the industrial discharge of mercury generated near panic in some parts of the country. In this case, federal officials quickly secured court-approved settlements that, within months, reduced mercury discharges from those sources to a small fraction of original levels. The federal government gradually developed a policy of using the criminal provisions of the Refuse Act in cases of irregular or occasional waste discharges, while employing the law's civil provisions to modify waste treatment processes.[69]

Congressman Henry Reuss of Wisconsin also helped publicize the potential value of the Refuse Act. Reuss, a Democrat from Milwaukee, established his reputation in Congress as an expert in the field of international finance, but he also compiled a strong record on civil rights and environmental issues. Reuss often joined with Democratic senator Philip Hart in calling for stronger pollution enforcement in the Great Lakes region. Like his Democratic colleague John Dingell, Reuss could be merciless in his questioning of witnesses unlucky enough to appear before his subcommittee. Reuss's flair for publicity and confrontational style sometimes irritated his fellow legislators, but they respected his intellect and integrity.[70]

In February 1970, a staff report prepared for Representative Reuss's Subcommittee on Conservation and Natural Resources called on the Corps of Engineers to prosecute polluters under the Refuse Act. In order to test the effectiveness of the Refuse Act, Reuss turned a list of 270 industrial dischargers over to the U.S. attorneys for eastern and western Wisconsin for action under the law. The Justice Department brought criminal charges against only four companies on the list but won convictions in each case. When Reuss applied for his half of the fines, as provided for in the law, the Justice Department balked but eventually paid the congressman his share. Reuss hoped that this financial incentive would encourage ordinary citizens to initiate Refuse Act enforcement. Reuss was impatient with the reluctance of the Justice Department to bring suit under the Refuse Act. At one point, he attempted to file a civil action himself against two Milwaukee-area firms after the U.S. attorneys refused to take action, but the federal judge dismissed the cases, ruling that a private citizen could not bring suit under the 1899 law.[71]

The federal EPA began operation in the same month that President Nixon announced the establishment of a new Refuse Act discharger permit program. Although the latter initiative was designed ultimately to lessen the need for litigation and bring greater consistency and stability to federal enforcement, top officials at the EPA emphasized the use of formal enforcement actions to spur the pace of pollution abatement during the agency's first two years of existence. When William Ruckelshaus assumed the position of EPA administrator in December 1970, his legal background and orientation, along with his desire to quickly establish the new agency's credentials with the environmental movement, led to the reliance on lawsuits and other formal enforcement measures. In the field of water pollution, Ruckelshaus and the team of young lawyers that he brought with him to the EPA decided to abandon the use of enforcement conferences and to rely more on direct legal action to abate pollution. Dischargers would still be required to meet the abatement timetables established by the enforcement conferences.[72]

Shortly after becoming administrator, Ruckelshaus proposed that the Nixon administration implement an accelerated Great Lakes abatement program that would allocate more EPA personnel and resources to this region. According to Ruckelshaus, "the one area that stands for the environment and its degradation in the minds of the American people is the Great Lakes. I am asked what we are doing about it in every corner of the country." He also noted that together the eight states bordering the Great Lakes accounted for 180 electoral votes. Despite Ruckelshaus's urgings, Office of Management and Budget director George Shultz and Nixon aide John Ehrlichmann vetoed this proposal on budget grounds. Nonetheless, federal water pollution control officials continued to give Great Lakes water quality a high priority.[73]

When the EPA began operation, federal enforcement conference requirements and abatement timetables covered municipalities and industries discharging to Lake Superior, Lake Michigan, and Lake Erie (including the Detroit River). By this time, the dischargers' earlier "slippage" in meeting the conferences' abatement timetables had become even more pronounced. At the last meeting of the Lake Erie conference in June 1970, the FWPCA reported that 44 (roughly one-third) of the 130 industries that had been given schedules for implementing abatement programs had fallen behind their timetables; 43 had missed their final construction date and one was behind on an intermediate phase of the program; 38 plants were over one year behind schedule. Whereas 68 percent of the industries had been able to complete construction of their treatment facilities by their given deadline, it had been determined that almost one-fourth of the new facilities were providing inadequate treatment and so would need additional capital improvements.[74]

The record for municipalities was even worse. Only 28 out of the original 110 municipalities determined to need additional treatment facilities had completed their programs. A mere four of the remaining municipalities were meeting their original schedules, while 49 units were more than one year behind and 56 had missed their final construction date. The Lake Michigan enforcement conference experienced similar delays for both industries and cities, but the later start of this conference meant that final completion deadlines came later than in the Lake Erie Basin.[75]

Murray Stein, who remained assistant commissioner for Enforcement and Standards Compliance when the FWPCA became the EPA Water Quality Office, became more willing to threaten legal action against lagging dischargers, although he preferred to employ 180-day notices. Stein hoped these notices would be enough to force laggards to accelerate their abatement programs and would also serve notice to other dischargers that the federal government would not tolerate unreasonable delays.[76] Stein and other EPA officials in Washington soon found out, however, that regional personnel had little experience in preparing for litigation. Even in the case of 180-day notices, headquarters was reluctant to proceed unless there seemed a reasonable chance for success in court if a suit proved necessary. But in the Lake Erie Basin, Stein complained about inadequate information from the regional office on individual dischargers, resulting in time-consuming verification of facts and requests for additional information.[77]

Despite the initial start-up problems, the EPA, in conjunction with U.S. attorneys who continued to initiate Refuse Act suits on their own initiative, did increase formal enforcement action in the Great Lakes Basin dramatically during the agency's first two years. In one of Ruckelshaus's first public acts as administrator, he issued 180-day notices to the cities of Cleveland and Detroit. Other high-profile actions followed. During the next two years, the U.S. Justice Department filed Refuse Act suits in the Great Lakes Basin against such major corporations as Atlantic Richfield, Inland Steel, U.S. Steel, Allied Chemical, Reserve Mining, Republic Steel, and many others. In one action, EPA issued 180-day notices to just about every discharger in Wisconsin's heavily polluted Fox River valley.[78]

In some cases, information provided by local environmental groups sparked Justice Department suits. In the Calumet area, for example, Indiana Citizens Water Pollution Research, a local organization composed of scientists and other concerned citizens, supplied U.S. attorneys with information that led to Refuse Act suits against Mobil, DuPont, and U.S. Steel (Gary) plants. In Cleveland, the northeastern Ohio branch of the Sierra Club, one of the most active of the national environmental groups, also worked to spark federal suits against polluters. The northeastern Ohio Sierra Club's legal action committee, headed by Jerome S. Kalur, a young lawyer, assisted the FWPCA and the Justice Department in their investigation of area polluters. In several cases, Refuse Act suits against Cuyahoga River industries followed complaints by the Sierra Club to the Justice Department.[79]

Like their state counterparts, the U.S. attorneys could point to some notable successes. In October 1972, DuPont's East Chicago plant, which was on the receiving end of a Refuse Act civil suit in February 1971, entered into a consent decree that included a detailed timetable for achieving the best available control technology by the end of 1976. In December 1971, Jones and Laughlin Steel once again found itself the target of federal enforcement action when the Justice Department requested the U.S. District Court to issue an injunction against the steel firm under the Refuse Act to prevent it from discharging cyanides and other pollutants into the Cuyahoga River. Attorney general John Mitchell denied that complaints from the Sierra Club prompted the action against Jones and Laughlin, even though Kalur had sent a letter to Justice Department officials in September urging legal action against the company. Approximately one year later, Jones and Laughlin agreed to a consent decree that required the plant to install state-of-the-art controls by July 1, 1975.[80] As in the similar state actions, the agreements won by the federal lawyers were notable not so much for their abatement deadlines, which stretched into the middle of the decade, but for the degree of treatment that was to be attained at the end of the court-mandated timetables. More and more, regulators sought to impose treatment requirements that were based on what experts considered the highest degree of waste removal that was technologically feasible for a particular industry.[81]

Business leaders in the Great Lakes region responded with indignation to this new style of regulation that—in their eyes—featured capricious and uncoordinated enforcement action from different levels of government, continually tightening treatment requirements, and an unfair portrayal of industry as the major culprit behind all of America's environmental problems. According to Murray Stein, the industry representatives he dealt with around the country were most concerned about the element of uncertainty that had entered the water pollution control field.[82] In other words, would the treatment standards that they installed today be declared inadequate next year because of upgraded water quality standards? In the past, industry engineers expected that pollution control facilities would always require additions or replacement in the future. But the process had been accelerated rapidly, and firms now had to deal with the sometimes conflicting demands of state and federal agencies and U.S. and state attorneys, not to mention the requirements of aggressive local authorities such as the Chicago MSD. Company officials continued to be alarmed at the trend toward imposing high minimum levels of treatment on all dischargers regardless of local circumstances, or "treatment for treatment's sake," as U.S. Steel president Edgar B. Speer put it. Industry engineers and corporate executives also complained that most pollution problems called for technical solutions, but that the government regulatory agencies had become dominated by lawyers, who relied on litigation to force solutions to these complex problems.[83]

Business was correct about the increased role of lawyers in water pollution regulation, but evidence indicates that the formal enforcement action associated with their emergence was indeed effective in accelerating abatement activity at individual plants. During the early 1970s, the Council on Economic Priorities, a public interest research group, performed in-depth studies of pollution abatement at the major firms in both the iron and steel and the oil refining industries. The council found great variability in pollution control efforts at the different plants investigated, even within the same company. In general, mills and refineries subject to strong government enforcement action and pressure from local citizens groups made significant strides in reducing pollution, whereas plants that had escaped these pressures showed far less progress. The general environmental policy at a company was an important determinant of abatement progress, but the single most important factor in the pace and effectiveness of pollution reduction was the regional regulatory climate faced by individual plants.[84]

The Federal Permit Program

Environmentalists applauded the heightened enforcement activity of the federal government and its new willingness to take major pollution sources to court. But despite the favorable publicity generated, the frequent recourse to litigation presented some problems. For one thing, the rapid proliferation of uncoordinated litigation—including independent state action—and the potential vulnerability to legal action of practically all dischargers under the Refuse Act provisions created a climate of confusion and uncertainty regarding government policy and treatment requirements. In addition, the time-consuming and complex requirements of the American judicial process meant that individual cases could drag on for years if the defendants were willing to contest the charges. Finally, in a related point, litigating individual cases was no way to achieve the degree of standardization and equitable treatment requirements that federal officials believed desirable.[85]

The Refuse Act permit program, established in December 1970, was one attempt to deal with these problems. In addition to the above concerns, administration officials also wanted to protect the growing federal investment in municipal treatment plant construction. It was clear that whatever the final form of the new amendments to the federal Water Pollution Control Act now being considered in Congress, there would be a substantial increase in federal funding for the sewage treatment plant construction grant program. A General Accounting Office study, however, warned that additional federal funding in this area would be wasted unless industrial waste discharges were brought under more effective control.[86]

Under the permit program, all industrial dischargers were required to obtain a permit from the Army Corps of Engineers. EPA officials were to work closely with the Corps in determining whether permits should be issued or denied to individual companies, based on compliance with applicable water quality standards. Under the executive order creating the program, the EPA administrator also had the right to impose additional standards or requirements when deemed necessary. Companies denied permits were liable to prosecution under the terms of the Refuse Act. As a first step in the program, all industries discharging to navigable waters were to submit detailed information on the characteristics of their effluent.[87]

The Refuse Act permit program came under fire almost immediately from environmentalists and their allies in Congress. Critics scored the program for linking permit requirements to state water quality standards, which environmentalists viewed as generally inadequate. When EPA officials said they intended to review closely only those permit applications from firms discharging to interstate waters, environmentalist suspicions only deepened. Senator Hart's Senate Subcommittee on the Environment held hearings in February 1971 on the permit program, and Representative Reuss joined the proceedings. Both legislators harshly criticized the new program, especially its reliance on state agencies and existing water quality standards.[88] The executive director of BPI, Alexander Polikoff, echoed these views, arguing that the new program would "sound the death knell for Lake Michigan." Instead, Polikoff urged greater use of civil suits under the Refuse Act. This would allow courts to mandate the implementation of waste recycling control systems wherever this was technologically and economically feasible. The following month, Polikoff sent a public letter to Ruckelshaus. If a permit system was unavoidable, Polikoff wrote, then the general policy should be that no permits would be issued unless the industry in question agreed to begin using the best available control technology as soon as possible to reduce or eliminate its pollution.[89]

The question of treatment requirements was the key issue in the permit program. But a federal Supreme Court ruling in December 1971 effectively halted the program before it had a chance to get underway. At the time of the ruling, the Corps of Engineers had issued a total of twenty new permits to industries across the country. The ruling came in the case of a suit filed by Jerome S. Kalur and Donald Large against the Army Corps of Engineers. Kalur had previously been active as a member of the Sierra Club in the prosecution of industries discharging into the Cuyahoga River. The two plaintiffs frequently canoed on the Grand River, a Lake Erie tributary in northeastern Ohio, and they filed a class action suit on behalf of all recreational users of the river, challenging the Corps's authority to issue discharge permits to plants on the river.[90]

The federal judge ruled that the permit program was subject to the provisions of the National Environmental Policy Act, which required that environmental impact statements be prepared by federal agencies prior to actions that could affect the environment. In other words, the Corps would be required to prepare an environmental impact statement for each permit issued under the Refuse Act. Given the tremendous number of permit applications, this additional requirement would be overwhelming. No more permits were issued, but EPA personnel continued to process the applications already received in anticipation that a similar program would be incorporated into the new federal legislation pending in Congress.[91]

Looking back on his first term of service as EPA administrator, William Ruckelshaus described the federal agency's relations with the state governments as "terrible." This description certainly held true in the field of water pollution, where state regulatory officials complained constantly about "the increasing duplication of activities, enormous requests for detailed information, and other [federal] activities which not only do not contribute to the control of water pollution but also often actually hamper ongoing state efforts." In the area of enforcement, even those state engineers with many years of service behind them were able to change with the times, to some extent, and become more accepting of federal enforcement action, provided state officials were kept well informed and—ideally—consulted beforehand. Even in Ohio, chief sanitary engineer George Eagle at one point asked Murray Stein to consider further federal enforcement action against some Ohio industries and municipalities in the Lake Erie Basin that were falling behind in their abatement programs.[92] The Refuse Act permit program, however, was another matter entirely. State officials in the Great Lakes states and across the country were unified in their condemnation of this program as a wasteful duplication of existing state programs that placed most of the responsibility for pollution regulation in the hands of federal officials. In the eyes of Perry Miller, executive secretary of the Indiana SPCB, and in those of his counterparts in the other Great Lakes states, the increasing federal role in regulatory decision-making was turning the state officials into "little more than errand boys" for the EPA.[93]

The concerns of state regulators had little influence on Congress. In October 1972, Congress passed the Water Pollution Control Act Amendments, which included a permit program specifically exempted from environmental impact statement requirements. The new permit program also included municipal dischargers under its provisions. Unlike previous legislation, the 1972 amendments clearly placed ultimate authority for water pollution control in the hands of the federal government.[94]

Big City Woes

The resentment state officials felt toward the federal program in part reflected their disappointment at the federal government's failure to fully fund the authorizations approved for the sewage treatment plant construction grant program. This was especially true for states like Michigan and Wisconsin that had followed New York's lead and begun prefinancing at least part of the federal share that each approved project was entitled to. New York's situation did not inspire confidence. By the beginning of 1972, the federal government's financial contribution to the state's Pure Waters construction program amounted to 7 percent of the total costs incurred by the local governments and the state.[95] Since the construction of municipal waste treatment plants was so dependent on federal and state financial aid, the municipal compliance rate with government-mandated abatement schedules was often far worse than that of industry. The situation was particularly difficult for the largest cities in the Great Lakes Basin, which had to contend with a declining tax base, a growing demand for city services, and a host of other chronic urban problems that appeared overwhelming by the early 1970s.[96] Unfortunately for Great Lakes water quality, the great bulk of municipal pollution was concentrated in these large urban areas. In the Lake Erie region, the difficulties experienced by the cities of Cleveland and Detroit illustrated the complex pollution control problems of large urban areas.

In William Ruckelshaus's first enforcement action as head of the EPA, he appeared before the National League of Cities a week after his formal confirmation and announced that the EPA was issuing 180-day notices to Atlanta, Cleveland, and Detroit for violation of water quality standards. This action helped establish the EPA's reputation as a tough enforcer of antipollution laws, but the state and local officials involved viewed the violation notices as arbitrary and unfair. Ralph Purdy of the MWRC thought the situation ironic, since he blamed Detroit's inability to meet its federally approved abatement timetable on the federal government's failure to provide the promised construction grant funds.[97] Cleveland mayor Carl Stokes found his city's predicament even more frustrating. Because of the Ohio WPCB's desire to spread federal grant money among as many communities as possible, as of April 1970, Cleveland had not received one dime of federal money for sewage treatment plant construction. The previous year, state authorities ranked Cleveland twenty-fifth in the priority listing for federal construction grant funds. Since Ohio's own construction grant program was designed to match Washington's contribution for individual projects, significant state funding was also unavailable.[98]

City officials in Cleveland and Detroit also faced rising construction cost estimates. The burgeoning inflation of the period was an important factor. Gerald Remus of the Detroit sewer authority estimated that inflation alone had raised the original construction cost estimate for the needed improvements from $104 to $159 million between 1966 and 1969. Planners had also underestimated the total costs involved in upgrading the sprawling, antiquated sewer delivery systems. This was especially true for Cleveland, where much of the problem stemmed from the inability of the city's sewer system to handle the growing effluent contributed by connected suburbs. Detroit was in a much better position than Cleveland, since the former city had established a centralized metropolitan authority years earlier to coordinate suburban connections to the Detroit system. Finally, when EPA officials met with representatives from the two cities after the 180-day notices they used the opportunity—in typical fashion—to gain commitments for higher levels of treatment than had been originally planned.[99]

By the summer of 1971, each city reached an agreement with the EPA to spend over $500 million to upgrade its sewage treatment system to a level considered acceptable by federal officials. Under the agreement, Detroit was to achieve 90 percent phosphorus removal by the end of 1975 and 90 percent BOD removal by the summer of 1976. Cleveland was to complete all needed facilities at its three treatment plants by September 1975 and extend new trunk sewers to thirty-two suburban communities by the end of 1977. Only about half of Cleveland's expenditures would be eligible for a federal grant contribution, since the balance of the needed work did not involve treatment facilities or interceptor sewers.[100]

The federal agreement with Cleveland immediately ran into difficulties, with the major problem being the inability of the city and its suburbs—which contributed 30–40 percent of the total waste entering the system—to agree on a plan for apportioning the costs and providing better coordination of the system. The stalemate dragged on for months, despite the issuance of 180-day notices to all the suburbs and a court-imposed building ban on both Cleveland and the suburbs. A Cuyahoga County Court of Appeals judge finally resolved the issue, with the backing of the EPA, when he gained agreement from all the local governments involved to create the Cleveland Regional Sewer District. Cleveland officials initially had control of the district, but management authority was designed to shift to the suburbs as the county's population continued to move outward from the city.[101]

The late 1960s and early 1970s were years of upheaval in American society, as citizens with a variety of agendas questioned once accepted values and sought to reform a political-economic system that they viewed as corrupt and unjust. Mainstream political leaders attempted to respond to these challenges in such disparate areas as foreign policy and racial and gender equality by reforming existing institutions. In the field of environmental protection, activists and their political allies focused not just on evidence of declining environmental quality, but on the established methods of regulating activity that affected the environment. Nixon administration officials created the EPA to increase federal effectiveness in pollution control, and William Ruckelshaus adopted a regulatory strategy aimed at gaining the support of environmentalists. In the Great Lakes region, federal officials continued to be the most responsive to the demands for more aggressive enforcement action and higher levels of water quality. But even at the state level, the desire of elected officials to respond to the concerns of the organized public—along with pressure from Washington—led to the creation of new agencies with an expanded capability to deal with environmental problems and the greater use of formal enforcement measures.

During the years that government officials, environmentalists, and industry executives were grappling with the general problem of waste discharge control in the Great Lakes region, they also faced the challenge of formulating a responsible policy for dealing with the pollution threats posed by toxic substances, detergent phosphates, and the large volume of waste heat discharged by lakeshore power plants. In these cases, regulators had to make decisions with major economic consequences that were based on incomplete and sometimes conflicting scientific evidence. In addition, growing awareness about the prevalence of toxic pollutants in the Great Lakes Basin brought public health concerns to the forefront of pollution control policy once again.

CHAPTER FIVE

The Burden of Proof

Pollution Control and Scientific Uncertainty

P ublic health concerns reemerged in the late 1960s as a central feature in efforts to control Great Lakes water pollution. But instead of battling pathogens that threatened epidemic disease, regulatory authorities now found themselves dealing with an array of toxic compounds whose long-term effect on the human beings exposed to them was often unclear but potentially deadly. Lack of experience with the effects of these compounds made it extremely difficult to determine what levels of concentration, if any, were acceptable. In addition, placing controls on toxic pollutants forced regulators to take a broader approach to pollution control, since many of these substances did not enter waters from the end of a discharge pipe. The publicity given to toxic substances as a result of fish bans, scientific studies, and the statements of environmentalists only increased demands for more effective regulation in the basin.

At the same time that state and federal regulatory officials were struggling to develop effective policies for controlling toxic pollutants, two controversies began that also raised questions about the correct regulatory policy to pursue in the face of incomplete scientific evidence about the effects of pollutants. Uncertainty about the link between detergent phosphates and eutrophication and between waste heat and environmental degradation made it very difficult to adopt and maintain a consistent regulatory policy. The choices faced by regulators in these two cases were even more complex, in one sense, since there was no direct threat to human health involved. Without the specter of disease and death, it was tougher to defend policies that imposed large economic costs on the basis of incomplete evidence.

The negative effects of thermal discharges from power plants were potentially grave, but not well established. In addition, scientists were most concerned about dangers stemming from the cumulative buildup of waste heat in a body of water, but the "threshold value" where serious harm would take place was impossible to place with any certainty. Because neither side in the debate over thermal pollution control in Lake Michigan could say with certainty when significant damage would begin to occur, or what the effect on the entire lake would be, the decision about the need for strict thermal discharge controls came down to a value judgment about the proper course of action when faced with uncertainty. Should the burden of proof lie with those who wished to alter the lake and possibly cause irreparable damage, or upon those who wished to impose major costs that would ultimately fall on the public for environmental controls that might be unnecessary? In the case of detergent phosphates, the most relevant part of the scientific community achieved a rough consensus about their culpability in lake eutrophication. However, the potential health dangers of various phosphate substitutes injected a new element of uncertainty into the decision-making process that slowed or halted the regulation of detergent phosphates at different levels of government.

In the controversies over detergent phosphates and waste heat, the increasing willingness—and ability—of interest groups to pursue their agendas outside the usual administrative sphere of water pollution control both contributed to the breakdown of cooperative pragmatism and reflected the ongoing changes in the regulatory system. The behind-the-scenes lobbying of the detergent industry continued the informal contact between government and business that was a central feature of cooperative pragmatism. But when these efforts proved unsuccessful, the industry also adopted a policy of challenging local and state regulatory laws in the courts. Environmental groups pressing for stringent control of thermal discharges also sought relief through other channels; in this case, the licensing hearings of the Atomic Energy Commission (AEC), which offered ample opportunity to pressure utility companies directly by dragging out the proceedings and delaying the issuance of operating licenses.

The policy debates over the regulation of detergent phosphate content and thermal discharges illustrate once again the importance of the American federal system in determining the outcomes of government regulatory policy. In the case of detergent pollution, municipal, county, and state governments in the Great Lakes region were willing to pass laws to limit detergent phosphate content, whereas the federal government pursued a more cautious course. In the case of Lake Michigan thermal pollution, however, the roles were reversed. The federal water pollution control agencies sought to impose stringent controls on newly constructed power plants, whereas the state authorities resisted the imposition of expensive cooling systems on public service companies.

Toxins Take Center Stage

Until the late 1960s, clean water advocates in the Great Lakes region expressed the most concern about the negative impact of pollution on recreation opportunities, natural beauty, and nonhuman life. Scientists and public health officials, however, took note of the dangers posed by the widespread use of various pesticides and the proliferation of new industrial chemical compounds. Rachel Carson's *Silent Spring*, published in 1962, was influential in raising public awareness about the potential dangers of the uncontrolled use of insecticides. But several years earlier, assistant surgeon general Mark Hollis warned of the public health implications of the changing character of the nation's waste effluent. The threat of waterborne infectious disease had been well under control for decades, Hollis noted, but he was deeply concerned about the potential health effects of many of the new substances being introduced into the environment—"especially from long-term exposure to low concentrations."[1]

When the Lake Michigan federal enforcement conference opened in the winter of 1968, the conferees decided to appoint a technical committee with representatives from the federal government and each state to develop recommendations for monitoring and controlling pesticide use in the basin. Pesticides drained into basin waters after being applied to the land. Testimony presented at the conference by Michigan state officials revealed the functional division on this issue that was common at both the state and national levels. Ralph MacMullen, director of the Michigan Department of Conservation, expressed great concern about the levels of DDT and dieldrin (another chlorinated hydrocarbon insecticide) being found in Lake Michigan fish. MacMullen urged that all persistent (or "hard") pesticides be either banned or used under the most stringent controls, with less persistent chemicals substituted whenever possible. But in a submitted statement, Dale Ball, director of the Michigan DOA, called for more research on the issue and warned that "public hysteria, aroused often by innuendo," would lead to the needless ban of all hard pesticides.[2]

The technical committee's report on pesticides, issued in November 1968, emphasized the lack of reliable information about pesticide use and its impact on Lake Michigan. The U.S. Bureau of Commercial fisheries, however, had analyzed approximately thirty species of Great Lakes fish in the last three years and found insecticides present in all species taken from Lake Michigan. Interestingly, fish taken from Lake Michigan revealed insecticide levels that were usually two to five times higher than concentrations in fish taken from the other Great Lakes. The Michigan Department of Conservation believed that DDT was probably responsible for the recent death of nearly one million coho salmon fry hatched in Lake Michigan as part of the department's salmon stocking program.[3] The technical committee concluded that other fish species in Lake Michigan probably faced similar reproductive hazards, while there was some evidence that widespread insecticide use was responsible for the declining populations of raptorial birds, such as eagles and osprey, in the basin. In one of its recommendations, the committee proposed that maximum concentration levels be established for various persistent pesticides and that these limits serve as objectives when trying to reduce pesticide levels in the basin.[4]

The question of what constituted a realistic threshold value for DDT concentrations in fish became important the following spring when the U.S. Food and Drug Administration (FDA) seized 22,000 pounds of Lake Michigan coho salmon food products with high DDT levels taken from Michigan waters. The FDA had been monitoring DDT levels in the salmon and decided to take action when sampling results increased from 8–9 parts per million (ppm) in January 1968 to 13–22 ppm in 1969 and the fish began to be sold across state lines. Environmentalists and some members of Congress were calling for a zero tolerance level, but the FDA decided to make 3.5 ppm the interim tolerance for DDT in fish until an expert panel could set a final tolerance. Michigan officials warned HEW secretary Peter Finch that such a low standard would cripple the Great Lakes fishing industry.[5]

When the governors of Minnesota, Wisconsin, Michigan, Illinois, and Indiana met at the request of Michigan governor William Milliken shortly after the FDA salmon confiscation, they petitioned the federal government not to establish interim or permanent pesticide levels until the Governors' Pesticides Conference had submitted recommendations. In August 1969, after technical committees established by the Governors' Pesticides Conference had completed their studies, the conference issued a general statement on the need to restrict the application of persistent pesticides and to phase out their use in favor of more benign alternatives. Michigan had taken the lead on this issue earlier in the year when the state Department of Agriculture cancelled all DDT registrations in the wake of the FDA salmon confiscation, except for some minor applications that would require only very small uses of the chemical. However, the governors also asked the FDA to establish a more "realistic" tolerance level for DDT in fish than the current 5 ppm (adjusted slightly from the previous 3.5 ppm) interim standard.[6]

The following year, the State of Michigan, which had a considerable investment in the coho salmon stocking program, formally petitioned the FDA to establish a tolerance of 15 ppm for DDT levels in salmon and 10 ppm for other fish. Michigan officials argued that there was no scientific evidence for the lower threshold, but FDA commissioner Charles Edwards maintained that public health concerns prevented him from raising the threshold. Michigan officials had the authority to establish their own safety value for salmon sold within the state's borders, but state authorities were reluctant to challenge the federal government on such a controversial issue. Officials in the Michigan DNR (formerly Conservation) also feared that such action would undermine Governor Milliken's strong position on controlling pesticide use.[7]

The movement to ban the use of DDT and other hard pesticides in the Great Lakes Basin and around the country is beyond the scope of this study. In the states bordering Lake Michigan, the revelations concerning pesticide concentrations in fish added fuel to the calls for a general ban on hard pesticides. In April 1969, Michigan became the first state in the country to institute a permanent ban on DDT use, and Wisconsin later followed in Michigan's footsteps. The federal government was also moving in this direction. In June 1972, the EPA approved a national interdiction on the use of DDT. The complete ban of some hard pesticides combined with greater restrictions on the use of others limited new pesticide inputs in the Great Lakes Basin. Unfortunately, regulatory officials would be dealing with the residues of previous use for some time to come. And pesticides were

not the only toxic compounds that water pollution control officials found themselves struggling to deal with as the decade of the 1970s unfolded.[8]

In 1969, a doctoral candidate at the University of Western Ontario alerted
Canadian authorities to the possible presence of dangerous concentrations of
mercury in Lake St. Clair fish. On March 24, 1970, Canadian officials, acting on
evidence of laboratory test results, seized 18,000 pounds of walleye caught in
Lake St. Clair and subsequently banned the sale and export of fish caught commercially within the Canadian waters of the lake. The Ontario Water Resources
Commission identified the Dow Chemical chlor-alkali plant at Sarnia, Ontario,
as the source of the mercury. This plant and other chlor-alkali plants produced
chlorine and caustic soda using mercury electrodes. The manufacturing process
also produced small amounts of mercury waste that were discharged with the
plant effluent.[9]

The public health danger of mercury buildup in the aquatic environment was
just becoming generally known in the United States at this time, primarily as a
result of unfolding revelations about the tragedy that had taken place in Minamata, Japan. Over a period of years, a vinyl chloride and acetaldehyne plant discharged large quantities of methylmercury into Minamata Bay on the island of
Kyusha. Unfortunately, fish from the bay served as a staple in the diet of local vil-
lagers. Japanese health authorities estimated that more than forty people died
from mercury ingestion between 1953 and 1960, when authorities placed restrictions on the plant's waste output. Many more persons suffered severe and
permanent damage to their nervous systems, including a number of congenital
cases. In the United States, business and government officials did not realize that
inorganic mercury—the kind used in the chlor-alkali production process—could
be biologically converted to methylmercury when discharged into an aquatic environment and then absorbed by fish either through ingestion of bottom organisms or directly through their gills.[10]

The federal food and drug authorities in both the United States and Canada
established 0.5 ppm as the safety threshold value for mercury in fish. Among the
states, there was little hesitation about what action to take when fish samples exceeded these levels, first in Lake St. Clair, then in Lake Erie, and then in other
parts of the country. Because of mercury's acute toxicity and the publicity given
to events in Japan, both federal and state authorities moved vigorously to control
the problem. In the weeks following the Canadian fish seizure, state officials in
the Lake Erie and Lake Ontario basins took a number of temporary actions to
protect public safety, including, at various times, bans on commercial fishing,
sampling and sometimes seizure of commercial fish catches, limiting sport fishing to a catch-and-release basis, and public advisories against eating fish from
contaminated areas. Together, these actions dealt another devastating blow to the
already struggling commercial fishing operations in these regions and also hurt
the lucrative outdoor recreation business along the lakeshores.[11]

State and federal authorities also moved aggressively against the industrial
sources of mercury that they were able to identify. In several cases plants were tem-

porarily shut down until the discharges could be stopped, and federal officials used the civil suit provision of the Refuse Act to gain injunctions against firms. These actions brought quick results. Within a year, the chlorine industry reduced the amount of mercury in its effluent by more than 95 percent.[12] The human health concerns brought to the fore by the pesticide and mercury issues, along with the increasing sophistication of chemical and monitoring analysis, forced regulatory officials to concentrate more of their efforts on toxic pollutants. Water pollution control agencies did not want to be caught by surprise again.

Polychlorinated biphenyls, or PCBs, soon emerged as another important toxic concern. PCBs had been used for decades in a variety of consumer and industrial applications, but their toxic characteristics alarmed public health officials, who were becoming aware of the compounds' ubiquity in the natural environment. Early in 1971, a Michigan official in the health department's Bureau of Laboratories warned a Governor Milliken aide that testing revealed concentrations of PCBs in coho salmon equal to or exceeding DDT levels. The toxicity of PCBs was supposed to be less than that of DDT, but the scientist was uneasy about the long-term effects of human exposure to PCBs. Later in the year, growing concerns about the danger posed by these PCB concentrations forced Governor Milliken to halt all commercial sales of the salmon. The federal government took steps to discontinue all future uses of PCBs, but their presence in existing products and persistence in the environment made these compounds a long-term problem for pollution control in the Great Lakes.[13]

The Reserve Mining Company and Lake Superior

Citizens living along the southern shores of the Great Lakes Basin did not have a monopoly on water quality problems, as developments in the Lake Superior Basin demonstrated. Lake Superior was the cleanest of the Great Lakes, owing primarily to the lack of population and development along its shorelines. Thus, when evidence began to mount concerning the degradation of Lake Superior water quality, environmentalists urged quick and decisive action to prevent the largest of the Great Lakes from going the way of its sister lakes.[14] The Lake Superior situation was unique in several respects, relative to circumstances in Lakes Michigan, Erie, and Ontario. First, attention quickly focused on only one source of pollution—the Reserve Mining Company's daily discharge of more than 67,000 tons of waste from its taconite processing operation on the shore of Lake Superior in Silver Bay, Minnesota. Taconite was a low-grade form of iron ore. Second, the Reserve processing plant and the company's mining operation almost fifty miles inland employed approximately three thousand people, making Reserve Mining by far the single largest employer in what had been an economically depressed area. Thus, the Reserve case assumed tremendous significance for every person in the region.

In addition, the opponents of Reserve's discharge would not be satisfied with anything less than onshore disposal of the "tailings"—the waste rock left over after the processing of the taconite. This made a compromise in the case much more difficult to achieve than in most industrial water pollution control cases. Finally, Reserve Mining's fervent contention that the taconite tailings were inert and caused no harm whatsoever to the lake gave this case a different flavor, since government scientists had to first prove the harmfulness of the tailing discharges before an abatement order could be issued. The question of harm took on a more sinister connotation when evidence emerged late in 1972 that tailing fibers entering municipal water supplies could have the same effect as asbestos fibers on persons who swallowed them.

Reserve Mining, jointly owned by Republic Steel and Armco Steel, had been the first company to begin processing taconite. Reserve utilized a method for processing raw taconite into superior-grade iron ore pellets that had been developed by scientists at the University of Minnesota. Their work could not have been more timely, since the high-grade iron ore in Minnesota's Mesabi Range had begun to play out by the middle decades of the twentieth century. Reserve Mining encountered little opposition when the company applied for a state permit in 1947 to dispose of taconite tailings in Lake Superior. The company also obtained a dumping permit from the Army Corps of Engineers without much difficulty. The processing plant at Silver Bay began operation in 1955.

A State Health Department study carried out in 1956–1957 raised serious questions about the movement of the tailings and their impact on the environment, but Minnesota authorities took no action, and the findings were not publicized. The real controversy began a decade later when a scientist at the FWPCA's National Water Quality Laboratory in Duluth conducted studies on his own initiative that indicated considerable tailing movement in the lake, a connection between the tailings and accelerated algae growth, and a possible impact on fish populations. The FWPCA downplayed these findings, and the scientist left federal service in 1967. Aside from Reserve's importance to the regional economy, the firm was located in the congressional district of John Blatnik, Democratic chairman of the powerful River and Harbors Subcommittee of the House Public Works Committee. Blatnik had been closely involved in authoring all of the post-1945 federal water pollution control legislation, but he was also a longtime booster of the taconite industry.

Concerns about Reserve's dumping came to a head in 1968 when Reserve requested renewal of its dumping permit from the Corps of Engineers. When some officials in the Bureau of Sport Fisheries and Wildlife and the FWPCA asked for further study before a new permit was issued, the Corps referred the matter to the Interior Department for study. The subsequent Taconite Study Group report concluded that a significant volume of the tailings was carried away from the immediate discharge area and that the tailings discharge caused turbidity, violated various water quality standards, stimulated algae growth, and adversely affected bottom fauna. The Taconite Study Group recommended eventual onshore disposal of the tailings. Reserve Mining vigorously disputed all these assertions. After an Interior official leaked the report to the *Minneapolis Tribune* in mid-January 1969, Blatnik, Secretary Udall, the director of the Minnesota Pollution Control Agency, and other state and federal officials downplayed the report, saying that it was just an unofficial draft and never should have been released. President Nixon's new team at Interior also adopted this line.

Secretary Udall, however, was not prepared to shrug off the matter. His friend Senator Gaylord Nelson of Wisconsin had been pushing for a federal enforcement conference on Lake Superior pollution for several years and the Taconite Study Group report offered considerable evidence that the tailings were crossing state lines and causing pollution. On January 19, 1969 (the next-to-last working day of the Johnson administration), Udall summoned an interstate enforcement conference on Lake Superior, his announcement explaining that the conference would be exploratory and fact-finding in nature. The events that followed moved gradually but inexorably against the interests of Reserve Mining, as mounting scientific evidence and a growing public clamor for action forced government officials into direct conflict with the company.

Federal and state conferees met officially for two sessions in 1969 and another two in 1970. During this period, federal technicians essentially confirmed the findings of the Taconite Study Group report and presented further evidence of harmful effects from the tailing discharges. Reserve officials continued to dispute the evidence and stalled on presenting a plan for corrective action, as Murray Stein had requested. At the January 1971 session of the conference, Reserve finally presented an abatement plan that called for adding an organic flocculent to the tailings and sending them through a deep pipe to the lake bottom, where they would remain. Stein, who faced pressure from his superiors to bring this case to a resolution, viewed the plan favorably, but the firm opposition of the newly elected governors of Minnesota and Wisconsin scotched any chance for quick approval of the plan. Instead, Stein and the other conferees appointed a technical committee to study the plan.

When the conference reconvened in April, the technical committee rejected the deep pipe proposal. The next day, David Dominick, who was serving as chairman of the conference, announced that he would recommend to William Ruckelshaus that the EPA issue Reserve Mining a 180-day notice. (Dominick recalled later that he was tired of the company's intransigence and had come to view the case as a national symbol of corporate disregard for the public interest.) The state conferees endorsed Dominick's action and then Grant Merritt—no longer a citizen environmentalist but head of the Minnesota Pollution Control Agency—proposed that the conference also adopt a recommendation for land disposal of the tailings. The conferees adopted this proposal unanimously.

Negotiations between federal officials and Reserve Mining subsequent to the 180-day notice failed to produce agreement. Meanwhile, political pressures to take Reserve to court mounted, as the intense lobbying efforts of environmentalists resulted in appeals for action from a number of area congressional representatives and senators. In January 1972, Ruckelshaus asked the Justice Department to bring suit against Reserve, and the following month Justice filed suit under the Refuse Act. Reserve had plenty of reason to fight the case. A federally sponsored study made after the suit was filed estimated that it would cost Reserve just under $190 million to rework its processing system and dispose of the tailings on land.

The efforts to force Reserve to cease discharging to Lake Superior had been driven by the desire to protect lake water quality. But the case took on a completely different flavor when federal scientists raised concerns about the similarity between taconite tailings fibers and asbestos fibers. The presence of the former in local drinking water supplies eventually forced top EPA officials to issue a public statement in June 1973 warning of the potential risk from ingesting these fibers and recommending that alternative sources of water be provided for young children. When the federal trial began in August 1973, it quickly became clear that this public health issue had displaced the previous environmental concerns. In this respect, the Reserve Mining case reflected the increased focus on environmental threats to human health that characterized the environmental movement after 1970.

Pressure Builds for a Ban on Detergent Phosphates

By the mid-1960s, government officials responsible for protecting water quality in the Great Lakes Basin viewed accelerated eutrophication as their greatest problem, especially in Lake Erie and Lake Ontario. Knowledge about the eutrophication process was incomplete, but a growing body of scientific evidence convinced many regulators that phosphorus input reduction was the most practical means of slowing eutrophication. At that time, however, methods for removing phosphorus at municipal treatment plants were not fully developed. Because phosphates used in household laundry detergents accounted for a large percentage of the phosphorus in sewage, some interested officials believed that eliminating phosphates from detergents was a logical course of action.

Phosphates—chemical compounds that contain the element phosphorus—were used in detergents to soften water and increase the detergents' cleaning power. The introduction of phosphates played a critical role in the growth of the detergent industry after World War II. Phosphates softened water and prevented dirt and other particles from being redeposited on clothes or the washing machine, thus dramatically improving the cleaning performance of the surfactant. Not only did detergents with phosphates clean noticeably better than soap, they were also less expensive.[15] Three firms dominated the market for finished goods in the detergent market: industry leader Procter and Gamble, Lever Brothers, and Colgate-Palmolive. Each of the Big Three belonged to the Soap and Detergent Association, the industry's trade association. Large chemical firms such as Dow and Monsanto that supplied raw materials for the industry were also members. In 1969, the membership of the Soap and Detergent Association represented over 90 percent of the soap and detergent production in the United States.[16]

Prior to the emergence of lake eutrophication as a policy issue, the soap and detergent industry resolved a less serious environmental problem involving its detergent products. The manner in which this problem was resolved foreshadowed the later controversy over detergent phosphates. The detergent formulations then in use were persisting after municipal sewage treatment, resulting in coatings of suds on some rivers and other bodies of water. The detergent manufacturers initially maintained that their products were not responsible for the foaming problem, but the industry succeeded eventually in developing a biodegradable surfactant, and in 1965 the Big Three and other producers voluntarily switched to the use of the new surfactant.[17]

This episode was significant to the debate over detergent phosphates because of the different lessons learned by the parties involved. Representative Henry Reuss, Senator Gaylord Nelson, and other Democratic members of Congress sponsored bills that would have required detergent manufacturers to begin using a biodegradable surfactant. Reuss and Nelson believed that it was the specter of such legislation that forced the detergent makers to change the formulation of their products. Later, when the relationship between detergent phosphates and lake eutrophication became a widely publicized environmental controversy, Reuss and other environmental advocates sought to duplicate what they perceived as their earlier success in the detergent foaming problem. Industry executives and federal government officials, however, viewed the resolution of the foaming problem as an example of how voluntary action by enlightened business leaders could resolve a technical problem with a minimum of government interference and red tape.[18]

When the technical committee established by the Lake Erie federal enforcement conference to study the eutrophication problem urged the detergent industry to develop a substitute for phosphates, Interior secretary Stewart Udall and industry executives established the Joint Industry-Government Task Force on Eutrophication. Charles Bueltman, technical director of the Soap and Detergent Association, served as the group's chair. The joint task force devoted most of its efforts over the next few years to the development of the Provisional Algal Assay Procedure (PAAP) test. This test, once perfected, would be used to evaluate the impact of any given compound on algae growth. The joint task force also engaged in research on phosphorus removal at sewage treatment plants.[19]

The call for the elimination of phosphates from detergents took on new urgency in late 1969 when the two bi-national boards appointed by the International Joint Commission to investigate pollution in Lake Erie, Lake Ontario, and the international section of the St. Lawrence River issued their final report. The IJC investigation could be traced back to June 1956 when Governor Averill Harriman of New York asked the U.S. State Department to make a formal reference to the IJC on Lake Erie water pollution. Other federal officials became involved, as well as governors of other states. Eventually, the State Department proposed to the Canadian government that a joint reference be made to the IJC concerning Lake Erie, Lake Ontario, and that part of the St. Lawrence River shared by the two nations.[20] The Canadian government failed to give formal approval to the reference until 1964. In October of that year, the two governments requested that the IJC undertake a study of pollution in the specified waters in order to determine whether pollution from one side of the border was damaging health and property on the other side, in violation of the 1909 Boundary Waters Treaty. If this was the case, the IJC was to determine the extent and causes of the pollution and recommend remedial action.[21]

The final IJC report emphasized the importance of reducing phosphorus content in the lakes as a means to slow and eventually reverse eutrophication. The advisory boards recommended that wastewater treatment plants achieve 80 percent phosphorus removal by 1972 in the Lake Erie Basin and by 1975 in the Lake Ontario Basin. In addition, the report called for detergent phosphate content to be reduced immediately to minimum practical levels, with an ultimate goal of completely replacing detergent phosphates with less harmful substitutes by 1972 in these areas. The boards estimated that phosphate-based detergents were responsible for 70 percent and 50 percent of the phosphorus in American and Canadian municipal waste, respectively. According to the report, eliminating phosphates from detergents would not only reduce phosphorus loading to the lakes but would save hundreds of millions of dollars annually that would otherwise be spent on chemicals used in phosphorus removal and sludge disposal.[22]

The IJC report and its recommendation that phosphates be removed from detergents were the subject of hearings held in December 1969 by Representative Reuss's House Subcommittee on Conservation and Natural Resources. Earlier in the year, Reuss introduced a bill in the House proposing to make illegal after June 30, 1971, the sale or manufacture of any detergent containing phosphorus.[23] Reuss viewed the detergent manufacturers as adversaries who would have to be coerced into taking action to benefit the public. The Milwaukee congressman began his hearings by stating that, if the detergent manufacturers did not reduce the amount of phosphates in their products, they should be made to share the cost of phosphorus removal at municipal treatment plants. Reuss believed that the soap and detergent industry had effectively captured control of the Joint Industry-Government Task Force on Eutrophication, and he ridiculed the group for devoting most of its energy to the development of the PAAP test instead of working directly on the development of a phosphate substitute.[24]

Soap and Detergent Association vice president and technical director Charles Bueltman, accompanied by representatives from the Big Three detergent manufacturers, provided the industry's response to Reuss's attack. Their presentation contained the basic arguments that would, with some adjustments, form the core of the industry's defense of detergent phosphates for the next decade. Bueltman and his allies emphasized the scientific uncertainty surrounding the issue of lake eutrophication. Bueltman strongly disputed the theory that the elimination of detergent phosphates would help solve the eutrophication problem, claiming that the available technical evidence did not support this contention. He called for more studies and cited several recent articles from scientific journals, the authors of which questioned the direct link between phosphates and excessive algae growth. These scientists argued that too many other factors were involved in the growth of algae to assume that a reduction in phosphorus would result in a proportionate reduction in algae.[25]

Bueltman and the others also argued that even if phosphorus was the critical or limiting element in the eutrophication process, reducing or eliminating detergent phosphates in the Great Lakes Basin would be ineffective, since large amounts of phosphorus would continue to enter the lakes from other sources—especially runoff from fertilized land. The industry representatives charged that in the case of Lake Erie, even if the recommended phosphorus reductions from municipal and industrial waste sources could be achieved, there would still be from other sources three to four times the amount of phosphorus necessary to produce nuisance algae conditions.[26] Bueltman assured the subcommittee that, even though the detergent makers questioned the link between phosphates and eutrophication, they were working as hard as possible on finding a substitute. The industry representatives stressed that research efforts aimed at finding a replacement for phosphates were time-consuming and complex because a substitute would have to be safe, effective, and readily available in large quantities and would have to exhibit no adverse effect on water quality. Bueltman mentioned one possible substitute, nitrilotriacetic acid (NTA).[27]

The Nixon Administration: Voluntarism or Legislation?

Reuss's hearings came at a time when officials in the Nixon administration were actively working to incorporate environmental protection issues into President Nixon's agenda. Administration staff members were not shy about asserting federal authority in this area and also did not want to see the Democratic Congress receive all the credit for national environmental initiatives. Direct interference with product formulations, however, would certainly not go over well with big business, an important Nixon constituency. To complicate matters, one of Nixon's top aides, perennial Washington insider Bryce Harlow, had worked as a lobbyist for Procter and Gamble before joining the Nixon staff. Moreover, Procter and Gamble chairman of the board Neil McElroy had served as secretary of Defense when Nixon had been vice president. Throughout 1970, members of the administration wavered between the options of continued reliance on the good faith of the detergent industry and regulation of detergent content.[28]

At the beginning of 1970, the White House staff tentatively planned to include a proposal for legislation to restrict detergent phosphate use as part of the president's upcoming environment message to Congress. The proposed law would gradually reduce the allowable phosphate content in detergents until a complete ban went into effect at the end of 1972. But when Nixon aide John Whitaker sent a memo to Harlow describing the plan, a horrified Harlow scrawled a note of protest on the memo, explaining that "I *absolutely* oppose this draconian stunt by *this* administration. It befits Kennedy, or Johnson, and I'm sure would delight Muskie, Nelson, Hart etc.—but how you can reconcile this with Nixon escapes me utterly." Harlow's actions to block this proposal are unknown, but during conversations just prior to the release of Nixon's environment message on February 10, the president agreed with the suggestion of Harlow and Whitaker that, rather than propose legislation to Congress, administration officials should try to "jawbone" the detergent makers into voluntarily reducing phosphate levels.[29]

Despite a series of high-level meetings between industry executives and cabinet secretaries and White House staff members, a voluntary agreement could not be reached. The detergent makers refused to make any voluntary reductions in phosphate levels and said that they would actually prefer legislation to voluntary action, although they opposed that course as well. In fact, the industry executives pledged to block any legislation that the administration introduced. Procter and Gamble president Howard Morgens threatened to take his case directly to the president, if necessary.[30] The staunch opposition of the major detergent manufacturers to any reduction in phosphates was based on two factors. First, as they made clear at Reuss's hearing in December, detergent executives believed that reducing phosphate content to the levels recommended by the administration would result in unacceptable cleaning performance for their products. Second, supplies of NTA, the only potential phosphate substitute available at that time, were insufficient to permit a complete substitution for phosphates. And the environmental effects and safety of NTA had not yet been completely established. The Big Three were not in complete unison, however. Executives at Lever Brothers and Colgate-Palmolive were more receptive to the idea of lowering phosphate content than Procter and Gamble officials, who refused to reduce phosphate levels without a corresponding addition of NTA. Since Procter and Gamble's detergents possessed the highest phosphate levels in the industry, the firm's major competitors might hope to gain from a government-mandated phosphate ceiling.[31]

In spite of the detergent makers' pleas for caution, the administration continued to pressure the industry to move more quickly. In a major reversal of Interior's previous emphasis on phosphorus reduction through waste treatment, FWPCA regional director Francis Mayo's testimony at an IJC hearing in Rochester in February 1970 supported immediately reducing detergent phosphate levels and eventually banning phosphates in detergents. Later in the year, the Department of Interior issued a public list of detergent products and their phosphate contents. David Dominick, commissioner of the renamed Federal Water Quality Administration (FWQA), denied that the administration was encouraging consumers to boycott those products with a high phosphate content. But Dominick appeared to contradict himself when he explained that the list had been released "to clear up any confusion as to the phosphate content in detergents and to serve as a general standard for the housewife for shopping in the supermarket."[32]

Within the administration, Russell Train used his position as chairman of the president's Council on Environmental Quality (CEQ) to press for federal legislation regulating detergent phosphate content. Train had enjoyed a distinguished public career before President Nixon appointed him undersecretary of the Interior in the new administration. The patrician Train first became interested in environmental issues during safaris to Africa and subsequently founded the African Wildlife Leadership Foundation. Appointed a U.S. Tax Court judge by President Eisenhower, Train resigned in 1965 to become president of the Conservation Foundation. When the National Environmental Policy Act created the CEQ, Nixon made Train the first chairman of the council in early 1970. In its early years, the CEQ played an important role in advising the president and developing administration policy on the environment.[33]

Train sought legislation that would authorize the secretary of Interior to regulate constituent levels in detergents. Ideally, the industry would then cooperate with the secretary and his staff in developing a schedule for gradually phasing out phosphates. Train believed that the industry had exaggerated the importance of high phosphate content and that, in most parts of the country, phosphate levels could be lowered without significantly reducing cleaning effectiveness. At this time, phosphate levels in detergents were usually expressed as percentage STPP (sodium tripolyphosphate), the chemical form of phosphate most widely used in detergent formulations. Phosphate content in the most popular brands of laundry detergent ranged from just under 35 percent to almost 60 percent. Train initially proposed that regulation should first place a ceiling of 25 percent on detergent phosphate content, but he later adjusted this to 35 percent. According to LeRoy Hurlburt of Colgate-Palmolive, detergents with a 35 percent phosphate content would provide adequate cleaning for 92 percent of the American population (the remaining 8 percent lived in hard water areas that required a higher phosphate content to maintain reasonable cleaning effectiveness).[34] It later became common to describe phosphate content using percentage of elemental phosphorus, which was roughly four times lower than percentage STPP. Thus, many of the local and state laws that initially limited detergent phosphate content before completely banning the compound, adopted a maximum level of 8.7 percent phosphorus or 34.4 percent phosphate.

Train hoped that government regulation would stimulate greater production of NTA, the phosphate substitute. He downplayed NTA safety concerns, arguing that responsibility for establishing the safety of NTA should rest with industry and that the federal government should not "be precluded from taking steps to protect the environment until *it* proves a substitute is acceptable." By June 1970, Train's legislative proposal had gained the support of Interior Secretary Hickel, Whitaker, John Ehrlichman, and other administrative officials. The major opponent was Commerce secretary Maurice Stans, who urged that no action be taken until further research was completed on the safety of NTA and the effects of the new phosphate-free detergents. Subsequent events justified Stans's caution.[35]

Government pressure to eliminate phosphates in detergents was successful in stimulating the industry's interest in NTA. By the end of 1970, Procter and

Gamble was adding NTA to a number of the company's brands with an accompanying reduction in phosphate content, while Lever Brothers had introduced liquid All, which entirely replaced phosphates with NTA. A Department of Commerce study estimated in the summer of 1970 that, given projected increases in NTA production, the amount of phosphates used by the Big Three detergent manufacturers would be cut in half by the end of 1972, without any government regulation.[36] But as production of NTA increased steadily, safety questions about the compound's use began to emerge. A Procter and Gamble rat-feeding experiment found that massive doses of NTA produced kidney disease in the rodents. The company downplayed the results, citing the extremely high dosage levels. The compound received more publicity when prominent Harvard Medical School pathologist Samuel Epstein warned that as NTA broke down in sewage it could possibly combine with other chemicals to form cancer-causing compounds.[37]

When government tests revealed that high doses of NTA combined with mercury and cadmium caused fetal abnormalities in rats, surgeon general Jesse Steinfeld and newly appointed EPA administrator William Ruckelshaus hurriedly convened a private meeting with industry representatives on December 17, 1970. Citing concerns about possible danger to pregnant women, Steinfeld and Ruckelshaus asked the industry to voluntarily discontinue the use of NTA, although the firms would be allowed to use up existing stocks of the material. The industry had little choice but to agree, since the government planned to release the findings to the public. Procter and Gamble, which had over $150 million of the compound on order, questioned the validity of the rat tests but complied along with the other firms. In a speech the following year, Commerce Secretary Stans used the NTA fiasco as an example of the problems that arose when government tried to impose hasty solutions on complex technical problems.[38]

Local and State Bans and Safety Concerns

Publicity about the eutrophication problem and the role of detergent phosphates set the stage for grassroots action in the Great Lakes region. In Buffalo in April 1970, a group of women formed Housewives to End Pollution in response to the IJC's findings on Great Lakes water pollution. The new organization began a public information campaign on the issue and investigated the problem in detail. In May, Housewives to End Pollution, threatening pressure tactics against Buffalo-area food stores and the detergent manufacturers, forced the Soap and Detergent Association to supply the group with data on the phosphate content of different detergent products. Housewives to End Pollution then posted large signs listing the phosphate content of different products in the detergent aisles at Buffalo-area grocery stores. Members of the group later acted as advisors to the Erie County (Buffalo) council when members of that body passed a phosphate restriction law.[39]

Legislation regulating detergent phosphate levels in the Great Lakes region first appeared at the municipal level and then spread to higher levels of government. Since this was an issue of product content and not regulation of discharger activity, laws restricting phosphate content in detergents provided an excellent opportunity for legislators to take a strong stance in the name of the environment without—in most cases—significantly affecting local economic interests. In October 1970, Chicago became the first city to act when its city council unanimously approved an ordinance first limiting and then banning the sale of phosphate detergents in the city. Akron passed a similar ordinance later that same year. In January 1971, the Detroit city council also adopted an ordinance limiting and then banning detergent phosphates.[40] Erie County followed suit in March. In April, Indiana became the first state to enact a phosphate control law, and New York followed with its own law in June. Like the others, the New York law first restricted the sale of detergents with a phosphorus content above a particular level, and then, after a certain time period, a total ban was placed on all detergents containing phosphates. The Erie County law provided for the earliest total ban, January 1, 1972. The Akron, Chicago, and Detroit bans were all scheduled to take effect after July 1, 1972. Indiana's ban was to begin on January 1, 1973, and New York's on July 1 that same year.[41]

These developments alarmed the detergent industry, to say the least. Representatives of the Big Three claimed that local ordinances limiting or banning detergent phosphates would not speed up the development of a viable substitute and would only create chaos as producers scrambled to meet the requirements of different local standards. Detergent industry officials lobbied hard against the new regulatory legislation—a councilman from Detroit described their pressure as "relentless"—but they had only partial success. Bills in Wisconsin and Pennsylvania did not become law, but the bans in New York, Chicago, and other areas represented a significant number of customers.[42] The detergent makers adopted legal action as a strategy to deal with those jurisdictions in which phosphate control legislation had been adopted. In January 1971, the Soap and Detergent Association brought suit in county court against the city of Akron to block enforcement of the city's phosphate restrictions. The parties reached an out of court settlement whereby the industry accepted a limit on phosphate content of 8.7 percent phosphorus and Akron agreed not to implement a total ban until industry representatives had a chance to air their views in a public hearing in the city. The Soap and Detergent Association also filed suits challenging the Detroit and Chicago bans, while Colgate-Palmolive took the same action against the Erie County ban.[43]

The industry did not confine itself to local ordinances. In June 1971, the Soap and Detergent Association filed suit in Indianapolis, seeking a permanent injunction against the enforcement of Indiana's new phosphate law. The association argued that the law was unconstitutional and that it would be impossible for the companies simultaneously to comply with the restrictions and to meet "accepted standards for human health and the environment." A Soap and Detergent

Association representative warned that the new law could result in the use of potentially dangerous substitutes that had not been fully tested. The Soap and Detergent Association case reflected the debate over phosphate substitutes that was taking place at the national level.[44]

The prohibition of NTA combined with public awareness of the eutrophication problem to create a niche in the market for other nonphosphate detergent products. The "dangerous substitutes" referred to in the Soap and Detergent Association's Indiana lawsuit were metasilicates and carbonates—commonly known as washing soda. A number of small manufacturers had begun to use these compounds as phosphate substitutes in detergents they made, and leading retail chains like Sears, Roebuck were marketing the new detergents under private labels. These products—given names such as Concern, Un-Polluter, and Phos-Free Detergent—were specifically aimed at capturing the business of environmentally aware consumers. By 1971, phosphate-free products had succeeded in capturing 10–15 percent of the detergent market.[45]

Representatives of the Big Three attacked the new phosphate-free products, charging that the highly caustic compounds contained in these detergents were potentially dangerous to human safety. Big Three representatives also questioned the cleaning effectiveness of the nonphosphate detergents. Lever Brothers and Colgate-Palmolive responded to the nonphosphate challenge and the proliferation of local and state laws by limiting phosphate content in all of their laundry detergents to 8.7 percent phosphorus. As of August 1, 1970, Canadian law made 8.7 percent phosphorus content the maximum level allowed in detergent products sold in Canada; industry experts considered this level the minimum necessary for adequate heavy-duty cleaning. Many of the local and state detergent phosphate laws also made 8.7 percent phosphorus content the temporary ceiling level prior to a complete ban. In contrast to its major rivals, Procter and Gamble continued to sell laundry detergents with much higher phosphate levels where legally permitted.[46]

Russell Train, CEQ chairman, initially viewed the new phosphate-free formulations as a possible breakthrough in the detergent-eutrophication problem. Mounting safety concerns, however, soon dampened his hopes. In February 1971, John Whitaker warned EPA administrator Ruckelshaus not to release a new listing of detergent brand phosphate levels. Such a list, Whitaker explained, would steer consumers toward high alkaline products that might be low in phosphates but were not as safe as the established brands, especially in homes with small children. The following month, the FDA seized 1,145 cases of Ecolo-G and another phosphate-free detergent brand—both manufactured by Ecology Corporation of America—after tests showed them to be toxic and harmful to skin and eyes upon contact. In April, Train testified before the Federal Trade Commission concerning proposed detergent labeling requirements that would spell out the harmful environmental effects of phosphates. He emphasized the importance of eliminating detergent phosphates but explained that concerns about the health hazards of phosphate substitutes mandated a cautious approach.[47]

Representatives of the Big Three pressured federal officials for an official announcement that made explicit government concerns about the phosphate-free detergents. In August 1971, a fifteen-month-old Connecticut girl died after swallowing a quantity of a phosphate-free detergent marketed by Arm and Hammer. On September 15, 1971, Train, Ruckelshaus, Surgeon General Steinfeld, and Charles Edwards, head of the FDA, issued a joint statement on the safety of caustic substitutes for phosphates at a Washington press conference. The federal officials labeled the health hazards of the caustic substitutes "a cause for serious concern" and strongly urged states and localities "to reconsider laws and policies which unduly restrict the use of phosphates in detergents." When asked, Steinfeld said that "my advice to the housewife is to use phosphate detergents. They are the safest things in terms of health." The joint statement said that the EPA would attack the eutrophication problem by working with states and municipalities to improve sewage treatment facilities that discharged to those lakes where phosphorus was the critical element in algae growth.[48]

The joint statement created considerable controversy. Representative Reuss and other environmentalists condemned the administration for selling out to the major detergent producers. Administration officials responsible for environmental policy attempted to clarify their position. Appearing at the annual convention of the Society of American Foresters in Cleveland, Ruckelshaus told a local reporter that there was no reason for housewives without small children to return to phosphate detergents. He denied he had ever recommended that communities on Lake Erie should rescind their phosphate bans. The decision was up to each community, the EPA chief said. In October, the subcommittees of both Representative Reuss and Senator Hart conducted hearings on the matter.[49]

One result of the Washington press conference was the introduction of a new detergent phosphate control bill by Senator Robert Griffin, a Republican from Michigan. The proposed detergent control act would limit the amount of phosphorus in detergents to 8.7 percent by weight and provide for federal preemption of state or local laws. Griffin considered the bill an excellent compromise. The use of reduced phosphate detergents would reduce the phosphorus input to the nation's lakes, while avoiding the hazardous potential of the widespread use of caustic phosphate substitutes. And the bill's preemptive clause would greatly simplify the marketing and production problems of the major detergent manufacturers. Not surprisingly, given their firms' adoption of a maximum 8.7 percent phosphorus content, representatives from Lever Brothers and Colgate-Palmolive strongly supported the national standardization of detergent phosphate regulations.[50]

Ruckelshaus testified that the administration opposed the Griffin bill because it did not believe the nationwide regulatory program involved in the bill was necessary, since phosphorus-driven eutrophication was not a problem in most parts of the country. The spokesman from Procter and Gamble shared the administration's emphasis on installing phosphorus removal processes at municipal treatment plants. Not only would this result in the removal of detergent phosphates from municipal effluent, but also phosphorus from other sources, as well as other plant nutrients found in sewage.[51] Advocates of detergent phosphate restrictions like Representative Reuss argued that the advanced treatment required to remove phosphorus from municipal sewage was very expensive and would take time to implement. And, in the end, the American taxpayer and financially strapped municipalities would end up footing the bill for these improvements. But some who had supported a phosphate ban began to register doubts. A spokesperson from Housewives to End Pollution noted the confusion caused in western New York by the September 15 joint statement. While Housewives to End Pollution continued to support Erie County's current 8.7 percent limit on detergent phosphorus content, the group called for further evaluation of the ban that was due to take effect at the beginning of 1972. A public hearing had been scheduled for November to consider the issue.[52]

Much of the debate at the hearings held by Hart and Reuss centered around the safety of the phosphate-free detergents. This proved to be anything but a clear-cut issue. Representative Reuss noted that tests recently conducted by the FDA on the hazardous qualities of detergents had resulted in a directive requiring many of the products—including some phosphate detergents—to begin carrying warning labels. Representatives from the Big Three questioned the practical validity of the FDA tests. According to the industry leaders, the phosphate-free detergents that offered the best cleaning tended to be the most alkaline, while the brands with low causticity were the least effective cleaners.[53] A member of Housewives to End Pollution gave credibility to the Big Three's criticisms of the nonphosphate detergents when she testified that "the introduction of phosphate-free detergents on the market has prompted consumer complaints of poor cleaning power, gritty-feeling clothes, residue remaining in laundry tubs and washing machines, burning hands and rashes upon contact, and excessively heavy packaging." In testimony before Senator Hart's subcommittee, Surgeon General Steinfeld qualified his remarks of September 15 by admitting that there were some phosphate detergents that were not completely safe, while there were some phosphate-free detergents that were not that caustic. In general, however, Steinfeld maintained, phosphate detergents were less hazardous to human health than those using substitute materials.[54]

Along with the product safety issue, the connection between detergent phosphates and lake eutrophication received much attention during the hearings. Although two of the Big Three detergent manufacturers supported the Griffin bill, all of them continued to question the culpability of detergent phosphates in lake eutrophication. They maintained that there was still no concrete scientific evidence that the elimination of detergent phosphates would be of any benefit, even in places like Lake Erie where eutrophication was a significant problem. Nevertheless, while the industry representatives referred to various scientific studies to back up their arguments, the consensus among limnologists (limnology is the scientific study of lakes) was that reduction of phosphorus would aid substantially in controlling eutrophication.[55]

An Uncertain Policy

Like Reuss's previous phosphate bill, Griffin's proposed legislation failed to make it out of committee. The debate over the safety of phosphate-free detergents and the accompanying publicity did have an impact on the detergent market, however, as sales of phosphate-free detergents began to decline sharply. The public safety debate combined with the availability of reduced phosphate detergents to soften phosphate control legislation in many parts of the country. In Michigan, Detroit and Flint had both passed ordinances that would eventually completely ban detergent phosphates, but a state law setting an 8.7 percent phosphorus limit superseded these bans before they could go into effect. When the Common Council of Detroit sought elimination of the preemption provision of the state law so that the city could proceed with its ban, Governor Milliken explained that a phosphate ban would force manufacturers to use hazardous substitutes. The Illinois Pollution Control Board also rejected a proposed statewide detergent ban, citing the health dangers of nonphosphate substitutes as one of the factors in its decision. Representatives from the Soap and Detergent Association, Procter and Gamble, Lever Brothers, and other firms tied to phosphate detergents appeared at board hearings on the issue, where they emphasized the health dangers of caustic phosphate substitutes.[56]

The detergent debate had important international ramifications, since controlling eutrophication was a major issue in the negotiations leading to the Great Lakes Water Quality Agreement of 1972. Because the majority of the Canadian population was concentrated along the shores of the Great Lakes, the problem of lake eutrophication generated even more popular concern—and greater political pressure—than in the United States. In Canada, an amendment regulating detergent phosphate content became part of a new water pollution control law enacted at the national level in 1970. In July 1970, the Canadian government issued regulations that limited the phosphate content of detergents sold in Canada to 8.7 percent phosphorus as of August 1. At the end of 1972, as planned, new regulations went into effect that limited detergent phosphate content to 2.2 percent phosphorus. Significantly, the Canadian government allowed detergent manufacturers to incorporate NTA into their products.[57]

Canadian officials realized that, due to the population disparity in the Great Lakes Basin, their measures would have little impact without significant phosphorus reduction from the American population. The Canadian government was thus disappointed when the Nixon administration retreated from efforts to reduce detergent phosphate content at the end of 1971.[58] The Nixon administration pledged to reduce phosphorus loadings to the Great Lakes through advanced waste treatment, but the unwillingness of the administration to regulate detergent phosphate content meant that the final agreement's phosphorus reduction objectives were significantly higher than the levels originally proposed by the IJC. Henry Reuss called the agreement "a victory for Procter and Gamble, but a disaster for Lake Erie."[59]

Reuss could take encouragement from events at the local and state level. The fact that many existing phosphate laws remained unchanged in the face of doubts about the safety of substitutes demonstrated the depth of environmental concern in the Great Lakes region, as well as the limitations of soap and detergent industry influence. The Big Three absorbed a major setback in December 1971 when a New York state judge ruled that the Erie County ban on detergent phosphates was legal. Colgate-Palmolive had challenged the law in court, but on the first of the year the Erie County ban went into effect, along with a similar ban in Dade County (Miami), Florida. These two markets had a combined population of almost 2.5 million.[60]

Lever Brothers became the first of the Big Three to break ranks when it began offering a phosphate-free detergent based on carbonate formulations in these two areas. In June 1972, Colgate-Palmolive announced that it would begin marketing phosphate-free detergents in areas where local ordinances prohibited the sale of phosphate-based detergents. Colgate-Palmolive also relied on a washing soda substitute in its new formulation, in this case sodium silicate compounds.[61] A federal court allowed the Chicago phosphate ban to go into effect on July 1, 1972, even though the Soap and Detergent Association had filed suit in the court challenging the law. Procter and Gamble later joined the suit, and the plaintiffs won a temporary victory when a federal judge ruled in the industry's favor and forced the city to suspend the ban. However, the U.S. Court of Appeals later overturned the ruling and the U.S. Supreme Court refused to hear the case.[62] In October 1972, Procter and Gamble announced that the firm would begin selling phosphate-free detergents on a trial basis in Chicago, Miami, and Buffalo. At the company's annual meeting of shareholders, board chairman Howard Morgens informed the audience that company scientists had developed a new surfactant system that would allow Procter and Gamble to begin selling laundry detergent with a phosphorus content of 6 percent. The new surfactant would also make Procter and Gamble's phosphate-free formulation safer than brands relying on washing soda substitutes. However, Morgens said that Procter and Gamble's formulation—like all other nonphosphate detergents currently available—would not offer "the proper level of cleaning effectiveness."[63]

In spite of intense industry lobbying, the Indiana and New York bans went ahead as scheduled. In both states, legislators introduced bills to delay the bans from taking effect, but they were unsuccessful. In New York, Department of Environmental Conservation Commissioner Diamond urged resistance to industry lobbying. In a report to Governor Rockefeller that called for implementation of the ban as scheduled, Diamond said the "housewives of this state have shown that they are more than willing to put up with a slight inconvenience in order to save our lakes."[64]

Thermal Pollution and the Growth of Nuclear Power

Thermal pollution, the heating of water bodies caused by the discharge of water that has been used for cooling purposes in industrial and power-generating plants, began to receive serious attention from government water pollution control officials in the early 1960s. As the decade progressed, evidence mounted concerning the harmful effects of thermal discharge on fish and the entire aquatic environment. Between 1962 and 1967, the FWPCA recorded ten cases in which waste heat from fossil fuel power plants caused significant fish kills. Even more troubling were the potential long-term effects of temperature increases on feeding, growth, reproduction, and other functions vital to sustaining healthy fish populations. Scientists also linked temperature increases to accelerated algae growth and the diminution of water's capacity to retain dissolved oxygen. These effects, however, were imperfectly understood and depended upon many local variables.[65]

Aside from the increased awareness brought about through scientific study, heightened concern about the effects of thermal discharge resulted from the prospect of a tremendous increase in the volume of waste heat discharged into the nation's waters. The electric power industry was responsible for approximately 80 percent of the total waste heat discharged by all industries. In postwar America, electrical power generation increased at an annual rate of approximately 8.5 percent. Experts foresaw continued growth in the demand for electric power, and the industry rushed to construct new generating plants to meet the expected need. Officials from the Federal Power Commission estimated that the additional power plants built to meet the growing demand would increase the power industry's total waste heat contribution to America's waters more than five times between 1970 and 1990.[66]

The expected transition of the power industry to the use of nuclear power plants was a critical factor in the projected thermal pollution increase, since nuclear plants discharged approximately 40–50 percent more heated water per unit of electrical production than conventional fossil fuel plants. During the mid-1960s, utility nuclear reactor orders—which had never exceeded more than three in one year—leaped from eight in 1965 to twenty-one in 1966, and then to twenty-seven in 1967. As reactor orders increased, so did the size of individual reactors. Average reactor capacity increased from 550 megawatts in 1963 to 850 megawatts in 1967. By the end of 1967, nuclear plants accounted for almost one-half of the total power plant capacity on order. Ironically, given later events, many executives in the power industry saw nuclear power as a means to meet America's growing energy needs while avoiding the pollution problems associated with the use of fossil fuels. As Glenn Seaborg, AEC chair, once put it, in comparison with coal "there can be no doubt that nuclear power comes out looking like Mr. Clean."[67]

An electric power plant used large quantities of water in its condenser to cool the hot steam from the plant's boiler after the steam's mechanical energy had been used to move the plant's turbine. After the steam returned to its original liquid state, the water was returned to the boiler. In the "once-through cooling" system then used by most American power plants, the cooling water (distinct from the boiling water) was returned to its original source at a point far enough away from the intake point to prevent immediate recirculation. In this system, the cooling water was returned to its source at an increased temperature of 10–30 degrees Fahrenheit.[68]

Methods to eliminate practically all thermal discharge had been in existence for several decades prior to the debate about the effects of waste heat on natural bodies of water. "Closed-cycle cooling" systems were originally developed in areas lacking adequate sources of cooling water. These systems either employed large cooling ponds (essentially man-made lakes) or cooling towers. Use of cooling ponds was restricted by their tremendous land requirements—typically two or three square miles for a 1,000-megawatt plant. The capital and operating costs of cooling towers depended on the characteristics of the particular plant in question, but the investment usually increased the total cost of the plant by 6–10 percent.[69] In systems using cooling towers, water leaving the condenser was sprayed onto a network of baffles and gradually cooled through the evaporation resulting from contact with a moving airstream. The cooled water was then recirculated to the condenser to begin the cycle again. The airstream inside the tower could be created either mechanically by large fans in the bottom of the tower or naturally by the rise of warm air in a much bigger tower. The capital cost of the larger natural draft tower was greater than that of a mechanical tower, but this difference was usually offset by the greater costs associated with maintaining and operating the mechanical system.[70]

As concern about thermal pollution mounted, environmentalists and their allies in government looked to closed-cycle cooling as a solution to the problem. In testimony before Congress in 1967, Interior secretary Stewart Udall said that instead of simply discharging heated water back to its source, the power industry was "going to have to invest in cooling towers." Unfortunately for Udall and his supporters, the FWPCA did not have the authority to impose effluent standards on dischargers. The problem would have to be attacked through cooperation with the states.[71]

The AEC possessed substantial regulatory authority over civilian nuclear reactors, but the agency denied any responsibility for regulating the waste heat discharges of the nuclear power plants it licensed, arguing that the agency's authority in environmental matters was limited to radiation hazards. Since the passage of the Atomic Energy Act of 1954, the AEC had worked with electric utility companies to support the development of nuclear power as a civilian energy source. This dual role as regulator and promoter came under increasing attack as public concern about the safety and environmental impact of nuclear power mounted in the late 1960s and early 1970s. Congressional environmentalists such as Representative John Dingell and Senator Edmund Muskie attacked the AEC for its position on thermal pollution, but the agency feared that placing strict controls on nuclear power plant thermal discharges would offer a competitive advantage to fossil fuel plants since they would not be subject to such requirements.[72]

Thermal Pollution in Lake Michigan

Of the Great Lakes, Lake Michigan would be most affected by the expansion of electric generating capacity planned for the 1970s. The existing electrical production capacity of power plants sited on the lake was expected almost to double during this decade with the added operations of seven nuclear power plants. All but one of the new nuclear plants were scheduled to be completed by 1973. The power plants already in operation were the largest users of cooling water from the lake. A federal study estimated that in 1968 40 billion British thermal units (BTUs) an hour were being injected into Lake Michigan, with electric-generating plants accounting for approximately 75 percent of the total waste heat.[73]

At the initial meeting of the Lake Michigan federal enforcement conference in early 1968, many of the conservation and civic groups that presented statements expressed particular concern about the pending buildup of waste heat discharges to the lake. Some urged that cooling towers be required for power plants located on the lakeshore. Secretary Udall, who was unable to attend the conference, also placed special emphasis on thermal pollution in his submitted statement. Udall noted the planned construction of nuclear plants along the lakeshore and argued that here was a situation in which pollution control authorities had a chance to take preventive action to eliminate a threat before significant damage took place. The conferees agreed subsequently to establish a committee to evaluate the potential threat to Lake Michigan from both thermal discharges and liquid radiation discharges. The committee was made up of representatives from the FW-PCA and the four state conferees.[74]

The committee presented its report at the next meeting of the conference in February 1969. The committee stressed the many gaps in knowledge about the effects of waste heat discharges on water quality. The report concluded that although the long-range effects of the temperature increases on the lake as a whole deserved consideration, there was little need for immediate concern. On the other hand, the potential local impact on aquatic life was an area that merited close examination. The committee noted evidence that fish and other aquatic organisms that served as their food supply were particularly sensitive to fluctuations in seasonal temperature during reproductive and juvenile stages. Abnormally high temperatures could upset the normal reproductive cycle of fish, thus limiting or preventing their production. There was also evidence that increased temperatures could accelerate the eutrophication of Lake Michigan.[75]

Other than recommending that all power plants on Lake Michigan meet their state's discharge requirements and water quality standards, the report avoided specific recommendations for action and instead called for further study of thermal pollution in the lake. Illinois, Indiana, and Wisconsin had water temperature standards for Lake Michigan, approved by the secretary of the Interior, which placed a limitation on the maximum temperature of different parts of the lake. Illinois and Wisconsin also limited increases above ambient temperature to no more than 5 degrees. Each state allowed a reasonable distance for "mixing" or assimilation of the discharge when measuring temperatures. Michigan's temperature standard had not been approved by the Interior secretary and was only a general statement that prohibited harm of any kind to the lake due to temperature changes. Murray Stein was disappointed in the committee's report, saying that recommendations for further study could go on forever. The conferees agreed to accept the report for consideration, postponing any substantive decisions on the issue.[76]

The problem of scientific uncertainty about the effects of thermal discharges prevented Stein and other federal officials from achieving the consensus that they desired on pollution abatement. When it came to controlling other forms of industrial pollution, interested parties might challenge the proposed abatement timetable or the degree of pollutant reduction that was feasible, but, except in unusual cases like Reserve Mining's taconite discharges, everyone agreed that the pollutant in question was undesirable and should be controlled. Complicating matters, the calls for stringent control of thermal discharges to Lake Michigan were based primarily on concerns about the effects of a future buildup in the total waste heat loading. Thus, proponents of stricter control had no existing significant environmental damage they could point to as concrete evidence of the negative effects of temperature increases. Conflicting scientific testimony also complicated the task of government pollution control officials. During the coming years, reputable scientists from the same disciplines would offer conflicting testimony about the dangers of thermal discharges in Lake Michigan.[77]

Since no party could say with complete certainty just how the additional waste heat would affect the lake ecosystem, the policy decisions had to be based on a value judgment. Should the burden of proof rest upon those who wished to alter the lake with additional thermal inputs? or upon those who wished to protect the status quo by imposing millions of dollars in new costs on power companies for safeguards that might not even be necessary? Environmental groups in the region favored the former course, whereas the utilities argued in favor of the latter policy. Government water pollution control officials, especially at the federal level, increasingly leaned toward the position of the environmentalists.

The Lake Michigan conferees may have put off a decision on thermal discharge control, but the issue continued to engage the attention of environmentalists. An article on thermal pollution of the Great Lakes in the January–February 1970 issue of *Environment* magazine charged that operation of the projected power plants on Lake Michigan using once-through cooling would be equivalent to heating the upstream Mississippi River 15 degrees and redirecting it through the lake. The thermal discharge policy ultimately adopted by the Lake Michigan conferees took on even greater significance for the region's power companies when Congress enacted the Water Quality Improvement Act of 1970 in March. One of the act's provisions required that any applicant for a federal license or permit had to obtain a certificate from the appropriate state agency certifying that the activity to be carried out under the federal authority would not violate state water quality standards.[78]

The Lake Michigan conference reconvened in the spring of 1970 and the conferees agreed to consider the thermal pollution problem at an executive session. Stein told the state representatives that it was important to have a single temperature standard for Lake Michigan and that they should aim for developing a consensus at the upcoming session. Clarence Klassen of Illinois and the other state officials were thus taken aback when Stein opened the May 7, 1970, executive session by reading a terse statement from Interior assistant secretary Carl Klein: "The minimum possible waste heat shall be added to the waters of Lake Michigan. In no event will heat discharges be permitted to exceed a one degree Fahrenheit rise over ambient at the point of discharge. This will preclude the need for mixing zones."[79] The state conferees questioned both the practicality and the necessity of the new policy, given the lack of any new evidence on thermal effects. Klassen said that not only would Commonwealth Edison's new Zion nuclear power plant be unlikely to meet the tougher standard but that other important industrial facilities on the lakeshore would also have trouble complying with the new temperature requirement. George Travers, director of environmental information at Commonwealth Edison, agreed with Klassen, adding that immediate enforcement of this "very, very extreme position" would force the utility to shut down two existing fossil fuel plants that discharged to the lake. Like other utility executives, Travers challenged the Interior Department to offer any concrete scientific evidence to support this arbitrary policy.[80]

The state conferees asked federal officials to provide the data necessary to justify the more stringent requirements. The conferees scheduled workshop sessions for September and October to consider technical evidence regarding the Lake Michigan temperature standards. Assistant Secretary Klein, whose abrasive and sometimes arrogant manner had alienated many officials both within and outside the FWPCA, resigned just prior to the opening of the technical sessions. Federal officials backed away from Klein's thermal discharge proposal: Interior secretary Walter Hickel explained that the one-degree limit had been entirely Klein's idea.[81] State and federal officials heard testimony from expert witnesses that lasted four days and filled over fifteen hundred pages of transcripts. The technical presentations raised more questions than they answered. Whereas scientists appearing on behalf of the Department of Interior and environmental groups testified on the need to halt waste discharges immediately in order to protect the lakes's health, the power industry presented its own lineup of distinguished scientists who challenged assertions about the harmful effects of allowing additional waste heat to enter the lake.[82]

When the conferees met in executive session at the end of October, they decided to appoint another technical committee to devise a "heat quota" system for limiting thermal discharges to Lake Michigan. Under this plan, limitations would be set on the total amount of heat—measured in BTUs—that could be discharged into a particular area of the lake from all sources, including municipal sewage plants and tributary streams. Once the heat quota had been reached in an area, any additional waste heat sources would either have to install a closed-cycle cooling system or locate in an area that had not yet reached its quota.[83] Federal officials made it known, off the record, that they expected the new plan to result in heat quota ceilings that would force both fossil fuel and nuclear power plants to install closed-cycle cooling systems. Environmentalists, however, condemned the conference for failing to take immediate action to protect Lake Michigan. In a press release, Congressman Dingell complained that after numerous meetings and studies the conferees could only agree to another study and a future meeting. Meanwhile, "in the absence of strong enforceable thermal pollution controls, these [power] plants will dump tremendous quantities of super-heated water into Lake Michigan and thereby set the stage for an ecological disaster."[84]

Environmentalist Intervention

While federal and state officials continued to study the problem, environmental groups in the Lake Michigan region sought to resolve the thermal pollution problem through other channels. Because of nuclear power's status as a highly regulated form of energy, nuclear power plants proved vulnerable to citizen pressure through the AEC's own regulatory process. Under the 1954 Atomic Energy Act, a new nuclear power station had to be licensed by the AEC Atomic Safety and Licensing Boards prior to both construction and initial operation. The AEC was required to hold a public hearing before a construction permit was granted, but the operating license was granted automatically after inspection requirements were satisfied unless someone requested and established the need for such a hearing. The nuclear power boosters on the congressional Joint Committee on Atomic Energy had included the dual licensing requirement in the 1954 act as a safeguard for the power industry. Design approval and the subsequent granting of a construction permit would provide formal assurance to the utility that its large capital investment would not go to waste. The earlier review process tended to make the granting of an operating license something of a formality. In fact, before the late 1960s the AEC and the Joint Committee on Atomic Energy viewed the public licensing hearings primarily as a means of building public support for nuclear power and reassuring the local citizenry of the plant's safety. But during the earlier period there was little public interest or participation in the licensing proceedings.[85]

Concern about nuclear power in the United States initially centered around questions of health and safety. However, until the 1960s, with a few exceptions, there was little public scrutiny or protest directed at the development of nuclear power. Increased public debate about the safety of nuclear power grew out of local battles over the siting of nuclear power plants. Reactor safety and radiation exposure were major concerns of early nuclear power opponents, but protection of the environment was also an issue in some of the siting disputes. In the early 1960s, for example, environmentalists were able to block the construction of a nuclear power plant at Bodega Head, a unique coastal site fifty miles north of San Francisco. In this case, opponents cited both aesthetic concerns and the proposed reactor's proximity to the San Andreas fault.[86]

Opposition to nuclear power accelerated in the late 1960s, fueled by the rapid expansion of nuclear power capacity around the country and books and articles that raised serious questions about the effects of nuclear power plants. Concerns about the environmental impact of nuclear power were intertwined with health and safety issues in the nuclear debate. Local opponents of nuclear power plant construction varied in their goals. Some hoped to prevent the construction or operation of the plant, whereas others sought to gain environmental or safety concessions from the electric utility that owned the plant.[87] In the states bordering Lake Michigan, the conflicts over the siting of nuclear plants were both an extension of the battle against water pollution and part of the controversy over nuclear power. The existing pollution problems of Lake Michigan made the prospect of increased thermal pollution a significant threat in the eyes of environmentalists. Critics of nuclear power in the Lake Michigan region often grouped their demands for reducing thermal pollution with other nuclear power concerns, but waste heat control remained a distinct issue. Unlike problems such as radiation discharge, nuclear power plant thermal pollution was entirely an environmental problem, since it posed no threat to human health or safety.

In the Lake Michigan Basin, environmental groups for the first time demonstrated the effectiveness of the AEC licensing hearings as a tool to extract environmental concessions from utilities when they intervened in the licensing hearings for Consumers Power Company's Palisades nuclear plant on the eastern shore of Lake Michigan. The Atomic Safety and Licensing Board hearings were legal proceedings; their trial-type procedures (pre-hearing conferences, elaborate discovery, cross-examination, and the like) could be utilized by knowledgeable participants to impose long and costly delays in the granting of a license or permit. Individuals or groups granted the right to intervene and take part in the proceedings were also able to appeal Atomic Safety and Licensing Board rulings to federal courts. Obviously, power companies were most vulnerable at the operating license stage of the licensing process because millions of dollars stood to be lost as a result of delays that kept the completed plant from going into operation.

When notice that an operating license was to be granted to the 800-megawatt Palisades reactor appeared in the *Federal Register* in March 1970, a coalition of local groups—the Michigan Steelhead and Salmon Fishermen's Association, Thermal Ecology Must Be Preserved (TEMP), Concerned Petitioning Citizens, and others—formally intervened and had their request for a public hearing granted. The Sierra Club and BPI later joined these groups as intervenors. When the hearings began in June, the intervenors sought additional controls for thermal and radiation discharge, but Consumers Power maintained that the existing facilities provided adequate protection in these areas. In addition, the Atomic Safety and Licensing Board ruled that the issue of thermal pollution was outside its domain and that it would only concern itself with safety and radiological concerns.[88]

The MWRC had already approved operation of the Palisades plant, even though plans called for cooling water to be returned to the lake 28 degrees warmer than ambient lake temperature. The Michigan agency's approval of such a discharge prompted a petition drive that gathered thirty-five thousand signatures. The Department of Interior also questioned the MWRC decision. At the licensing hearings a representative from the Department of Interior read a statement from Secretary Hickel that made clear the Department's intention to apply retroactively the thermal standards eventually agreed upon by the Lake Michigan enforcement conference. Hickel's statement said it was unlikely that the Palisades plant with its present design would meet any standard approved by the FWQA.[89]

Prevented from examining the plant's thermal discharge policy, lawyers for the intervenors focused on issues of plant safety. Using a variety of legal tactics, the intervenors dragged out the proceedings for the rest of the year. Finally, in December 1970, Consumers Power—which claimed it was losing $1 million a month and being forced to buy power from other utilities—agreed to negotiate with the environmentalists. In March 1971, Consumers Power and the intervenors signed an agreement under which the utility would install a cooling tower and modify its waste handling system so that radioactive discharges to Lake Michigan would be reduced to "essentially zero." In return, the intervenors agreed to the issuance of an operating license and the utility was allowed to begin operation of the plant while the new facilities were being constructed. Consumers Power vice president Russell C. Young continued to maintain that the new antipollution equipment was not needed, but he explained that an agreement was necessary to get the plant into operation. The utility put the total cost of the new equipment at $15 million, with two-thirds of that going toward the cooling tower. The additional safeguards would also mean an extra $3 million per year in operating costs.[90]

In the wake of its victory over Consumers Power, BPI sent copies of the Palisades agreement to every utility planning to operate a nuclear power plant on the shores of Lake Michigan, along with a suggestion that the utility follow suit in order to avoid the costly delays encountered by Consumers Power. The public interest group admitted that the issue of thermal discharge was outside the jurisdiction of the Atomic Safety and Licensing Board, but a BPI official said, "if the hearing goes on for a long time and explores all sorts of issues, the utility may get the idea that it pays to settle."[91] The possibility of losing millions of dollars while engaging in a long procedural battle with environmentalists had already convinced one utility to grant environmental concessions. Two weeks before the Palisades agreement, Northern Indiana Public Service Company became the first power company to officially announce plans to construct a cooling tower to prevent thermal discharge to Lake Michigan. Northern Indiana Public Service had not yet obtained a construction permit for its planned Bailly nuclear power plant in Indiana.[92]

Power companies in the western Lake Erie Basin also moved voluntarily to control thermal pollution. While the situation was not as serious as in Lake Michigan, a federal report at the June 1970 meeting of the Lake Erie federal enforcement conference noted the great increase in waste heat loading to western Lake Erie that could be expected from the construction of three power plants—two nuclear-fueled and one fossil-fueled—that were scheduled to begin operation on the lakeshore within the next five years. George Harlow, head of the FWQA Lake Erie office, called for closed-cycle cooling at all major heat sources.[93] The following month, Toledo Edison Company announced that the new Davis-Besse nuclear plant, which was jointly owned with Cleveland Electric

Illuminating Company, would make use of a cooling tower. Utility officials cited uncertainty about the temperature standards that would ultimately be adopted for Lake Erie. Local environmental groups had also been pushing for closed-cycle cooling at the plant. The bottom line was that it would be much cheaper to incorporate a cooling tower into the original design than to add it later. In May 1971, two months after the Palisades agreement, Detroit Edison reached an accord with representatives from BPI and the Sierra Club that called for the utility to spend $20 million for pollution-control equipment, including two cooling towers, at its Fermi nuclear power plant near Detroit.[94]

The decisions by Toledo Edison and Detroit Edison were part of a broader trend. Power company executives continued to question the need for closed-cycle cooling, but as historian Samuel Walker notes, "the costs of responding to litigation, enduring postponements in construction or operation of new plants, or suffering loss of public esteem were less tolerable than those of adding towers or ponds." Thus, by 1971, all utilities applying to the Atomic Safety and Licensing Board for a construction permit included closed-cycle cooling systems in their plans. The decision to adopt closed-cycle cooling was more difficult for operating plants or plants under construction, since adding cooling towers at this stage—usually referred to as backfitting—was much more expensive and disruptive. This was the case with most of the Lake Michigan nuclear plants.[95]

The Federal-State Split on Temperature Standards

When the Lake Michigan enforcement conference reconvened in March 1971 in Chicago, once again to address the issue of thermal standards, the power companies involved in the controversy anxiously awaited the conference's recommendations. A utility could hope to outmaneuver or outlast the intervenors in the AEC licensing process, but if the states were to adopt new thermal standards that mandated the addition of closed-cycle cooling systems, there would be little choice but to comply. Impatient environmentalists also waited for the outcome of the reconvened session, hoping that this time government officials would adopt a tough stance and stick with it.

The conference opened with the reading of a statement from EPA administrator William Ruckelshaus that called for the installation of closed-cycle cooling systems at all new power plants discharging to Lake Michigan. This included plants currently under construction and completed plants that had not begun operation. Ruckelshaus justified the policy by referring to increasing evidence that large volumes of waste heat—as would be built up in Lake Michigan under existing requirements—contributed to acceleration of the eutrophication process, seriously affected the health of aquatic life, and had other negative effects on water quality. The EPA head also urged the states to adopt clearly defined, uniform water temperature standards.[96]

When the conference's technical committee presented its report, the recommendations supported Ruckelshaus's policy, although the committee remained equivocal about the effects of thermal discharge on the lake. The committee's chair explained that the lack of conclusive data on the impact of thermal inputs rendered unworkable the heat quota plan suggested at the October 1970 meeting of the conference. The committee was, however, concerned about the local effects of thermal discharges and took the position that local ecological damage must be assumed unless proven otherwise. Therefore, the committee recommended that all power plants discharging to the lake be required to install closed-cycle cooling or other appropriate controls unless the plant in question could demonstrate that ecological damage would not occur from once-through cooling.[97] An EPA Region 5 official presented temperature standards to accompany Ruckelshaus's recommendations. Under the proposed standard, thermal discharges would not be allowed to raise water temperature more than three degrees above temperature maximums established for each month of the year or the current ambient temperature, whichever was lower. Each discharge would be allowed a 1,000-feet mixing zone. The EPA believed that existing power plants would either require no modifications or would have to modify their once-through cooling system discharge structure at minimal expense.[98]

Power industry representatives at the conference reacted very strongly against Ruckelshaus's policy statement, especially in light of the technical committee's report, which failed to present any significant new evidence on the effects of waste heat. The legal council for Indiana and Michigan Electric accused Ruckelshaus of caving in to environmental interest group pressure by making a purely political move divorced from technical considerations.[99] Environmental groups at the conference were, of course, pleased with Ruckelshaus's statement. The Lake Michigan Inter-League Group of the League of Women Voters conceded that research up to this point had failed to prove conclusively the harmful effects of thermal discharges on the lake. The league maintained, however, that it was better to pursue a cautious policy of limiting the discharge of any more waste heat until officials were certain about its effect.[100]

The most important response to the new federal proposal came from the state conferees. The officials representing Michigan, Indiana, and Wisconsin agreed to endorse the federal position, at least as a policy they could present to their state boards for consideration. Illinois, however, challenged the federal proposal and refused to concur in any conference recommendations that mandated closed-cycle cooling for power plants already under construction. David Currie, chairman of the Illinois Pollution Control Board, explained that the board's major concern was preventing further proliferation of major new heat sources along Lake Michigan. Thus, he fully supported a policy of requiring closed-cycle cooling for all future power plants on the lakeshore. But Currie did not believe there was sufficient evidence about harm from individual plants to justify the large expenditures necessary to backfit cooling towers on the Zion nuclear plant and the other plants in question. EPA officials were dismayed at Currie's stance. Regional administrator Mayo raised the possibility of imposing thermal discharge requirements through use of the Refuse Act permit system or legal action by other Lake Michigan states.[101]

Despite federal objections, the Illinois Pollution Control Board several months later officially adopted thermal standards that essentially duplicated the conference's recommended standards, except that Zion's two 1,100-megawatt reactors would be allowed to operate without cooling towers.[102] Currie's reasoning proved attractive to the members of the Michigan Water Resources Commission and the Wisconsin Natural Resources Board. (Since the only new power plant on Indiana's section of the Lake Michigan shoreline had already committed to a cooling tower, that state could adopt the federal standard with no qualms.) In September, the MWRC adopted thermal standards for the lake that placed a forty-two-month moratorium on the construction of new power plants that did not incorporate closed-cycle cooling while further studies on the problem were conducted. However, the MWRC exempted the Cook and Palisades nuclear plants from the moratorium and also ruled that mixing zones would be established on a case-by-case basis. In December, the Wisconsin Natural Resources Board also adopted a policy of nonproliferation that required closed-cycle cooling for all future power plants on the lake, but exempted the new Kewaunee and Point Beach nuclear plants. Like Michigan, the Wisconsin standards incorporated the three-degree rise limitation recommended by EPA but provided that mixing zones would be set individually, based on future study.[103]

Within the EPA, Francis Mayo argued that the agency should make vigorous use of all administrative and legal avenues available to it to push for the installation of closed-cycle cooling at the targeted plants, including intervention at the AEC licensing hearings. Top EPA officials were unwilling to take the latter action, but representatives from the agency did appear at the state temperature standards hearings to argue in favor of cooling towers at the nuclear plants. Mayo also took steps to prevent the Corps of Engineers from issuing Refuse Act permits to these power plants, but a court ruling placed the federal permit program on hold at the end of 1971.[104] By the spring of 1972, John Quarles, deputy administrator of the EPA, was urging Ruckelshaus to find a face-saving way for the agency to back down from its original recommendations and approve the standards adopted by Illinois, Michigan, and Wisconsin. Quarles pointed out that, if the EPA challenged the state standards either through the enforcement conference or through the convening of a standard-setting conference (as provided for under the Water Quality Act of 1965), the expected deadlock would lead to the convening of a hearing board. Given the composition of the board, which placed the EPA in a minority, they could expect the backfitting requirement to be voted down.[105]

At the same time that state and federal water pollution control officials were attempting to achieve a consensus on Lake Michigan thermal standards, area utilities and environmental groups continued their battle over this issue on other fronts. By the summer of 1972, officials from BPI had announced their intention to intervene in the operating license hearings for each of the lakeshore nuclear power plants that continued to resist the installation of cooling towers. The pattern of citizen intervention in the Atomic Safety and Licensing Board licensing process was being duplicated in other regions of the country.[106]

Power companies faced more delays in the wake of the landmark federal court decision in July 1971 in the case of *Calvert Cliffs Coordinating Committee vs. AEC.* This decision grew out of a citizen challenge to the construction of a nuclear power plant on the western shore of Chesapeake Bay, about thirty miles from Washington, D.C. The judge in the case ruled that the AEC had not been obeying the provisions of the National Environmental Protection Act. The court instructed the AEC to reorganize the agency's licensing procedures so that detailed study and consideration of the environmental impact of a proposed nuclear plant would have to be made before the issuance of construction permits or operating licenses. This would require the AEC to consider possible thermal pollution problems. The court's decision applied retroactively to all permits and licenses issued since January 1, 1970, the date the National Environmental Protection Act took effect. Hoping to restore the AEC's credibility with the public, the new chair of the AEC, James Schlesinger, decided against an appeal and announced that the agency would comply with the court's ruling. The most immediate result was that the AEC suspended licensing hearings for most of 1972 while Schlesinger and other AEC officials worked to overhaul the entire licensing process.[107]

The growing ability of scientists to monitor changes in the environment presented regulators with increased opportunities for progress. But these scientific and technical advances also created a more complicated policy-making environment. The incomplete nature of scientific knowledge concerning new pollutants such as PCBs and waste heat opened policy decisions to criticism from various parties with a stake in the outcome. Complicating matters, the toxic nature of many of these pollutants only heightened the emotional nature of the public debate over the best means for dealing with these problems. Sharp disputes over scientific evidence and appropriate levels of risk would remain a mainstay of water pollution control policy in the Great Lakes region and the rest of the nation.

Epilogue
The 1970s and Beyond

I n many ways the 1972 amendments to the Federal Water Pollution Control Act represented the culmination of the gradual movement away from cooperative pragmatism—in which state regulatory officials relied on informal cooperation with dischargers and possessed much administrative discretion—toward a new national system based on formal procedures, clearly defined statutory requirements, and a greater willingness by regulators to pursue formal legal action against dischargers. The efforts of citizen environmentalists and their allies in the media and government to create a new system of water pollution control aimed at uniformly high water quality had paid off. Despite these significant changes in regulatory policy, however, the characteristics of the old system did not completely fade away. Given the various avenues for appeal available to dischargers (both explicit in the new law and inherent in the American legal system) and the complexity of establishing numerical standards for designations such as "best practicable control technology currently available," administrative discretion and negotiation with affected parties were essential to the advancement of the program. In addition, while regulators sometimes used legal suits and other negative incentives against cities that were slow to upgrade treatment systems, the abatement of municipal pollution was closely tied to a federal construction grant program that remained inadequate despite increasing allocations. This fact, along with the financial limitations of declining cities, left regulatory officials with no choice but to grant delays and accept performance below statutory requirements.

State officials in the Great Lakes region may have been dismayed at their loss of autonomy under the 1972 law, but they continued to play a critical role in the regulatory process. Although the amendments placed final authority for administering the new permit system in the federal EPA, key provisions provided for the delegation of permit program authority to the states. By the mid-1970s, most of the states bordering the Great Lakes had been delegated this authority. While the state pollution control programs had improved greatly over prior years, they remained more attuned to the local and political implications of stringent enforcement. Eventually, despite well-publicized delays, pollution abatement efforts in the United States and Canada reduced dramatically the volume of pollutants flowing from the discharge pipes of industries and cities, bringing visible improvement to many parts of the Great Lakes Basin. This progress has continued through the present day, but the Great Lakes continue to face serious pollution problems, especially the persistence of toxic pollutants that appear to be even more harmful to all forms of life in the basin than originally feared.

The Great Lakes Water Quality Agreement

Prior to the enactment of the new national water quality law, mounting international pressures forced the United States to enter into a formal agreement with Canada that committed the two nations to take whatever steps were necessary to achieve specific water quality objectives in the Great Lakes Basin. In 1969, the two binational boards appointed by the IJC to investigate pollution in Lake Erie, Lake Ontario, and the international section of the St. Lawrence River issued their final report. The three-volume report contained much new scientific evidence, but only confirmed the obvious—pollution emanating from each side of the boundary was causing injury to health and property on the other side, and the United States was responsible for the lion's share of the total discharge volume. The report's major recommendations called for the implementation of accelerated phosphorus removal programs, the adoption of new water quality objectives by the IJC, and the establishment or strengthening of the control programs needed to achieve those objectives. The boards also advocated the creation of a new permanent bilateral agency to coordinate and monitor the new programs.[1]

Canadian officials sought a formal international agreement with the United States to carry out the IJC recommendations in all of the Great Lakes. Officials in the Nixon administration were concerned about the costs of such a commitment, especially after federal officials backed off on previous efforts to reduce or eliminate detergent phosphates because of the potential dangers of phosphate substitutes. This policy reversal meant that significant additional expenditures on municipal treatment would be needed if the United States was to meet the IJC's phosphorus reduction objectives by the proposed deadlines.[2]

Canadian officials continued to press Washington for an agreement. President Nixon and Prime Minister Pierre Trudeau finally signed the Great Lakes Water Quality Agreement at Ottawa on April 15, 1972. The agreement was a landmark in some respects, but its provisions fell short of what many in both countries had hoped for. The agreement established specific water quality objectives for a number of pollutants (though some were qualitative in nature). It was left to the individual governments, however, to implement the programs necessary to achieve these objectives. Significantly, such programs were only required to be "in process of implementation" by December 31, 1975. The agreement also enumerated yearly target phosphorus loadings to both Lake Erie and Lake Ontario through 1976, although these objectives were significantly higher than the levels originally proposed by the IJC. In addition, the agreement included two new references for investigation. First, a comprehensive study of pollution in Lake Superior and Lake Huron. And second, an investigation of pollution from nonpoint land use activities such as farming.[3] But probably the major significance of the Great Lakes Water Quality Agreement was that it established common objectives, including a nondegradation philosophy, for the entire basin and created a mechanism for the continuing evaluation and oversight of abatement progress in the entire Great Lakes system. As part of the latter thrust, the agreement granted the IJC new responsibilities for coordination, research, and dissemination of information, including the issuance of annual reports on water quality in the Great Lakes.[4]

Watershed: The 1972 Amendments

A close student of environmental policy has described the 1972 amendments to the Federal Water Pollution Control Act as the purest example of the "technology forcing" environmental protection laws enacted by Congress in the 1970s. This legislation—with its grand objective of ultimately eliminating all water pollution discharges and its provisions for establishing an array of exacting effluent limitations—"serves as an enduring monument to the American politician's belief in the possibilities of social engineering and to the political muscle of the environmental movement in the early 1970s." The new law, set forth in eighty-nine single-spaced pages of often highly technical language, was one of the most complicated pieces of legislation enacted in the U.S. Congress to that time. The meaning of key provisions of the amendments remained a topic of legislative and legal debate for years after the law's passage.[5]

Members of Congress, state and federal officials, and members of interest groups realized the significance of the new legislation. Prior to the enactment of the amendments in October 1972, these groups worked for a water pollution control bill that would meet their own particular agendas. By the early 1970s, American industry faced increasingly stringent environmental regulation from all levels of government. Business representatives complained about the confusion, uncertainty, and delays that resulted from overlapping government jurisdiction and blurred regulatory responsibilities. But in the area of water pollution control, at least, business did not welcome an expanded federal role as a way to bring stability to the situation. Instead, business representatives sought legislation that would clarify regulatory responsibility and leave as much authority as possible with the states, since industry officials could more effectively influence policy at the state level. Representatives for the major polluting industries remained wary about national effluent standards, once again arguing that individual discharger limitations should be geared to local conditions and needs.[6]

State officials were, if anything, even more critical of the legislative proposals that ultimately became the backbone of the 1972 amendments to the Federal Water Pollution Control Act. State water pollution control administrators in particular scored the discharge permit program, which put the states in the position of carrying out directives handed down from Washington. In a letter to Michigan governor William Milliken, Henry Diamond of the New York Department of Environmental Conservation complained that the proposals being considered paid little attention "to the experience of the states and their achievements." This national program, he argued, would render the states "agents of the Federal government."[7] Diamond was certainly correct about the limited attention given to state concerns in drafting the bill. Senator Edmund Muskie's Subcommittee on Air and Water Pollution developed the original bill that, with some revisions, later became the 1972 amendments. The staff members who worked on the bill generally scorned state water pollution control officials as being overly sensitive to the concerns of dischargers and hence too accepting of pollution. The staff members found common ground, however, with young environmental activists such as David Zwick, who had recently coauthored a study sponsored by Ralph Nader's Center for Study of Responsive Law that unmercifully criticized federal and especially state water pollution control programs. These activists helped shape the bill.[8]

Administration officials were ambivalent about the Muskie bill. While the permit program and some other key features had originated with Nixon officials, the sweeping water quality goals and strict deadlines of the legislation raised serious concerns about cost and feasibility. White House officials were able to get the more conservative House Public Works Committee to adopt a bill that altered many of the provisions in the Senate version. The final bill that emerged from the Senate-House conference retained the basic framework of the Muskie bill, but relaxed some of the deadlines and made other changes in line with administration sentiments. Still, the bill's price tag—$18 billion for construction grants and more than $24 billion for all expenditures—was too much for President Nixon. Despite advice from EPA administrator Ruckelshaus that he sign the bill, Nixon decided to veto the legislation, citing budget and inflation concerns. The next day, both houses overrode his veto by large margins.[9]

The main objectives of the 1972 amendments were unambiguous, if controversial. The ultimate goal was to eliminate completely the discharge of all pollutants to navigable waters by 1985. An interim goal was to render all navigable waters fit for fish and other aquatic life and for swimming and other forms of recreation by 1983. The means for achieving these goals were based on the conclusion that it was simply too difficult and complex to base regulatory efforts on case-by-case evaluations of the individual discharger's impact on local water quality. Instead, the new approach designated in the law applied minimum effluent limitations to different discharger categories, based on technological and economic feasibility, regardless of local receiving water conditions. When, however, these standard treatment requirements were not sufficient to attain the designated water quality standards for a particular area (which in turn had to satisfy national minimum requirements), dischargers would face more stringent treatment requirements.[10]

The national effluent treatment requirements were to be phased in gradually. Industrial dischargers had until July 1, 1977, to employ "the best practicable control technology currently available" (BPT), and until July 1, 1983, to employ "the best available [control] technology economically achievable" (BAT). Under the law, the EPA was responsible for determining the corresponding levels of treatment for each major industrial group. For their part, municipal dischargers were required to achieve secondary waste treatment by the 1977 deadline and "best practicable waste treatment technology" by the 1983 deadline. To make this feasible, Congress raised the federal contribution to 75 percent of the total cost of individual sewage treatment plants.[11]

The key instrument for implementing these provisions was a refurbished discharger permit program, known as the National Pollution Discharge Elimination System (NPDES), now expanded to include municipal treatment plants. Under the program, all industrial and municipal dischargers were required to obtain a permit that specified the amount and type of pollutants they were allowed to discharge, along with a schedule for implementing the remedial measures necessary to attain mandated treatment levels. The law gave the EPA the responsibility of administering the program, but the agency was encouraged to delegate this authority to individual states that met the requirements for taking over permit responsibility. Even in these cases, however, the EPA retained veto power over individual permits and could revoke the state authority over the program if federal officials deemed the state effort inadequate.[12]

The 1972 act represented a watershed in the control of water pollution in the United States. Many of its key features represented a culmination of regulatory trends that extended back into the previous decade. Most obviously, the federal government now possessed firm control over a regulatory program that had become truly national in its scope and structure. The new law also gave federal regulatory officials the authority to control directly the content of the effluent leaving the discharger's pipe, instead of having to work indirectly through water quality standards. Along these same lines, the revised federal law rejected the traditional doctrine of "reasonable use," or linking treatment requirements to the primary uses of receiving waters, and instead adopted a regulatory approach that aimed for the maximum degree of treatment feasible and uniform high water quality. Finally, in the area of enforcement the amendments eliminated the mandatory delays and limited federal jurisdiction of earlier laws. The 1972 law allowed the EPA or the state with permit-granting authority to issue cease-and-desist orders or to initiate suits against dischargers violating permit provisions after a thirty-day notice. Civil penalties could range up to $10,000 per day.

The New System: Litigation and Negotiation

Congress's desire to formalize the regulatory process and restrict administrative discretion led to a marked increase in litigation. Business contested the effluent guidelines and individual discharge permits that were needed to implement the NPDES program, while environmentalists tried to force regulatory authorities to follow the letter of the law. Prior to the July 1977 treatment compliance deadlines, the most important enforcement cases in the Great Lakes Basin concerned Refuse Act suits initiated before passage of the 1972 act and Illinois attorney general William Scott's continuing legal crusade against Calumet area steel mills. At the same time, however, close bargaining between government and business over the terms of the effluent guidelines and discharge permits also characterized this initial period of implementation.

On the municipal front, President Nixon had pledged in his veto message that if the water pollution bill became law he would use the spending discretion and flexibility that his aides believed the legislation conferred on the executive to limit construction grant funding. Nixon followed through on his threat and impounded $6 billion of the $11 billion authorized for construction grants in the fiscal years 1973 and 1974. Several years later, the Supreme Court ruled that the president did not have this authority, and President Gerald Ford released the balance of funds.[13]

As it turned out, President Nixon's impoundment of construction grant funds was not that significant in slowing municipal abatement progress because the 1972 amendments imposed a complex and burdensome array of new bureaucratic procedures on the grant application process, which brought the award of grants almost to a standstill. Congressional liberals' distrust of state and local officials and the former's accompanying desire to ensure that federal funds were spent wisely resulted in a cumbersome three-step review process with twenty-seven discrete decision points. To make matters worse, EPA officials failed to issue the required guidelines and regulations for many months and held up approval of state construction grant priority lists. In the fifteen-month period between the passage of the new law and the end of 1973, the federal government obligated only $35 million for new projects, with thirty-three states getting no money at all. Critics accused the EPA of intentionally stalling the construction grant administrative process in order to serve the administration's budget objectives. After congressional hearings in 1974 served as a forum for angry criticism of this bureaucratic quagmire, EPA officials revised and simplified the construction grant application and review process.[14]

The rapid rise in construction costs during the 1970s also undercut municipal progress. Between mid-1973 and the beginning of 1976, the cost of an average sewage treatment plant increased 50 percent. The major cities in the Great Lakes Basin were plagued by other problems that delayed progress toward abatement goals. These ranged from environmental impact statement requirements for new facilities at Cleveland and Syracuse (the latter stemming from a citizen lawsuit) to paralyzing contractor disputes at Cleveland. But the most frustrating problem for state and federal regulatory officials was the continuing low levels of operation and maintenance at many of these facilities. The science of sewage treatment was such that careful maintenance and operation was necessary for treatment facilities to operate at peak levels of efficiency. Unfortunately, financially strapped cities such as Detroit and Cleveland with sprawling facilities and sometimes antiquated equipment were particularly susceptible to this problem.[15]

The process of upgrading waste treatment at municipal plants at least had the virtue of a widely accepted initial target—secondary treatment. In the area of industrial waste discharge, EPA officials faced the daunting task of determining what constituted best practicable control technology for an array of industrial categories. The law required the EPA to issue these effluent guidelines by October 1973. The premise was that the regional offices and states with permit-granting authority would then use the guidelines in determining individual permit requirements. All industrial dischargers were to have permits by December 1974.[16] It soon became clear, however, that the authors of this scheme had vastly underestimated the complexity of developing uniform effluent guidelines for different industry groups that reflected the best control methods in use and yet were both technologically and economically achievable for most plants within the compliance timetable mandated by the law. The 1972 amendments specified 27 industrial categories for which the EPA was to establish effluent guidelines, but agency officials subsequently identified 180 industrial subcategories and 45 additional variances that required distinct effluent standards. The difficulty of the task was such that even the use of outside contractors did not enable the EPA to issue a single guideline by the October 1973 deadline.[17]

Industry actions also delayed issuance of the effluent guidelines. Along with an unwillingness to supply EPA personnel with information that would expedite the development of the effluent guidelines, individual firms and industry trade groups pursued a strategy of widespread litigation, formally challenging in court most of the proposed effluent guidelines. In general, these suits challenged the requirement that federal and state officials incorporate the specific limits in the guidelines into the individual permits. The industry briefs also argued that the 1972 act required the EPA to carry out cost-benefit analysis when considering abatement requirements for individual plants—not just for the industry groups as a whole, as EPA officials maintained. In addition, corporate attorneys charged that the guidelines were based on the performance of a few exemplary plants that did not necessarily reflect the conditions and problems at many other plants in the group.[18] Ironically, many of these suits came after comments from industry representatives had led to substantial revisions in the guidelines proposed initially. The federal courts ultimately ruled in favor of the EPA on the broad challenges to the agency's authority to establish specific numerical limits as part of the effluent guidelines and to incorporate these standards into individual permits. Industry was more successful in getting the EPA to adjust specific guidelines that were being challenged in court.[19]

A lawsuit filed by the National Resources Defense Council, an environmental group founded in 1970 that specialized in litigation and legislative lobbying, finally forced the Effluent Guidelines Division of the EPA to promulgate a number of guidelines in January 1974, with more following during the balance of the year under a schedule established by the U.S. district court for the District of Columbia. Meanwhile, top EPA officials decided that they had no choice but to begin issuing discharge permits if they hoped to meet the law's July 1977 treatment deadline. The Enforcement Office developed individual permits relying on data from the Refuse Act permit program applications and other sources. Thus, the great majority of major industrial dischargers received their permits before the related effluent guidelines were finalized. Confusing matters, the Enforcement Office produced interim guidance figures for twenty-two industrial categories that in many cases were more stringent than the actual guidelines finally promulgated. Such discrepancies provided fertile ground for permit appeals.[20]

The 1972 law granted dischargers the right to contest the terms of their NPDES permits through adjudicatory hearings. If unsatisfied with the administrative ruling, the discharger could appeal to the federal courts. The hearings could last anywhere from ten days to six months and court appeals could drag out the process even longer. The officers and attorneys of the major industrial dischargers in the Great Lakes Basin and throughout the country did not hesitate to contest permit terms when they believed it in their best interest. Most of the major steel mills in the region took this path, along with other large dischargers such as the Ford River Rouge complex.[21] State assumption of the permit program provided industry with a better chance to negotiate favorable adjustments, although the EPA retained a veto over the permits. Michigan became the first state on the Great Lakes to assume responsibility for administering the NPDES permit program in October 1973. By October 1975, all of the Great Lakes states except Illinois and Pennsylvania had been granted authority to administer the program.[22]

When Congress passed the 1972 amendments, John Quarles, deputy administrator of the EPA, promised a new era of swift, reliable enforcement that would ensure compliance with the clear, detailed abatement requirements contained in discharge permits. EPA legal enforcement action actually declined significantly during 1973 and 1974, however, in large part because the amendments directed that no discharger that had applied for a permit under the provisions of the new legislation could be prosecuted before December 31, 1974.[23] Suits previously initiated under the Refuse Act of 1899 were outside this mandate, and the EPA continued to pursue some of the Refuse Act suits that involved large industrial dischargers.

The most dramatic court case that fell into this category concerned Reserve Mining Company's discharge of taconite tailings into Lake Superior.[24] The threat of widespread cancer among the population from the ingestion of taconite tailing fibers made tenable drastic regulatory action that would not have been considered otherwise. In this case, state pollution control authorities in Minnesota, Wisconsin, and Michigan sought to close down the plant in order to stop the discharges. EPA officials, however, wavered in their support of such drastic action. Critics of the Nixon administration's caution in this matter pointed to the strong political support given to Nixon by Republic Steel and Armco Steel, Reserve's parent companies. Reserve Mining officers denied that the tailing fibers posed any threat to public health, but in March 1974 Reserve attorneys announced their willingness to negotiate a settlement based on land disposal of the tailings. Determining a proper land site and the necessary controls for both air and water pollution from the new operation proved complex, and the negotiations were contentious. At one point, federal judge Miles Lord ordered the plant to cease operations, but the Eighth Circuit Court of Appeals overruled him. Negotiations between Reserve Mining and the state of Minnesota over proper land disposal dragged on for years. After many court appeals by the parties involved, Reserve Mining finished construction of an inland tailings impoundment and ceased all discharges to Lake Superior in March 1980.[25]

In the Calumet region, Illinois attorney general William Scott continued to win legal settlements from area steel mills that included commitments from the firms to spend millions of dollars on remedial abatement programs. Scott and his staff decided early on to press for the installation of expensive waste recycling systems at the steel plants, pointing to the environmental damage that had resulted from trying to fine-tune effluent levels to take advantage of the receiving waters' assimilative capacity. Instead of the latter approach, Scott and his assistant attorneys pursued a policy of pollution regulation through lawsuits that forced dischargers to implement the highest level of waste treatment that was technologically feasible.[26] In this respect, Scott's objectives mirrored those of the federal water pollution control program after 1972. Through his suits, Scott was able to secure the installation of wastewater recycling systems at Inland Steel, Youngstown Sheet and Tube, and U.S. Steel's Gary and South Chicago Works.

Enforcement action under the NPDES permit program remained low-key even after most of the permits had been issued, in part because regulatory officials looked ahead to the July 1977 national compliance deadline as the point when laggards would face tough enforcement action. In the meantime, federal and state officials responded to most permit violations with administrative orders or more informal warning letters, which specified the violation and required compliance by a certain date. Many of the violations involved minor problems, such as failure to submit reports or to describe effluent characteristics adequately. In the case of municipalities, EPA officials adopted an informal policy of not taking enforcement action against cities and towns when their permit violations were the result of a lack of construction grant funds.[27]

The Importance of Federalism

The 1972 amendments to the Federal Water Pollution Control Act appeared to herald a new era of federal dominance in the field. However, during the 1970s the Great Lakes states demonstrated that they would still play a critical role in water quality policy. Eutrophication control was a good example of the continuing importance of federalism. Federal officials in the Great Lakes region used the success of the state detergent phosphate bans to support their argument that the federal government should reevaluate its position on this issue.

In July 1974, EPA regional administrator Francis Mayo passed along an Indiana State Board of Health report to Russell Train, now head of the EPA, that showed the dramatic reduction in the phosphorus content of Indiana sewage following the state's detergent phosphate ban. Mayo argued that Indiana's experience and similar results from other areas with bans suggested the need for a reappraisal of the EPA's position.[28] The IJC continued to push for a ban on detergent phosphates in the Great Lakes Basin. The commission noted the failure of major sewage treatment plants in the United States, such as those in Detroit and Cleveland, to achieve their prescribed targets for phosphorus removal.[29]

EPA officials took the Great Lakes Water Quality Agreement seriously and were unhappy with the failure of the U.S. government to meet the specified phosphorus reductions. When a new study by federal scientists concluded that the federal government's approach to phosphorus control was not sufficient to achieve the desired goals in the Great Lakes Basin, Train was forced to reconsider the need for eliminating phosphates from detergents. But probably the most important factor in the EPA's reevaluation of the phosphate issue was the experience in those areas that had implemented a phosphate ban. As expected, the phosphorus content in sewage had been reduced significantly. Most important, the problems predicted by the detergent industry had not materialized, at least not to the point that there was any kind of public outcry in these states and cities. There had been no rash of detergent-related injuries, no mass protests from housewives dissatisfied with cleaning performance.[30]

In August 1976, George Alexander, the new administrator for EPA Region 5, appeared before the MWRC in support of a proposed state ban on detergent phosphates. Alexander informed MWRC members that the EPA was now seeking such a ban in all the states bordering the Great Lakes. Alexander gave specific approval to the use of washing soda formulations and called for a reappraisal of the use of NTA, arguing that scientists had completely refuted previous allegations about the compound's safety.[31] In 1976, Minnesota became the third state in the Great Lakes Basin to ban detergent phosphates when the Pollution Control Agency board voted to implement such a ban as of January 1, 1977. The board decision came despite heavy industry opposition during public hearings and board meetings. As in the past, when detergent industry executives failed to achieve their objectives at the legislative and administrative level, they turned to litigation. A subsequent court ruling enjoined Minnesota from enforcing the ban, but the ban was later allowed to take effect.[32]

In Michigan, the failure of the Detroit treatment system to attain targeted phosphorus removal rates of 80 percent lent support to those advocating a state phosphate ban. Representatives from the soap and detergent industry offered their usual arguments against a ban, but proponents of the more stringent regulation gained a boost when Governor Milliken came out in support of a ban during his State of the State Message in January 1977.[33] The following month the Michigan Natural Resources Commission approved an administrative rule restricting detergent phosphate content to 0.5 percent phosphorus. The ruling went into effect later in the year despite a court challenge from the detergent industry. In 1978, Wisconsin governor Martin Schreiber signed into law a bill limiting detergent phosphate content to trace amounts. Thus, by the end of the decade, five of the eight Great Lakes states had banned the sale of detergents containing more than trace elements of phosphorus.[34]

In the debate over the appropriate policy for regulating thermal pollution in Lake Michigan, state and federal officials had also parted ways, with the former refusing to mandate the backfitting of expensive cooling towers on lakeshore nuclear power plants. Enactment of the 1972 amendments offered EPA officials a way out of the stalemate with their state counterparts. The new law did away with the enforcement conference mechanism, but on November 9, 1972, the former conferees to the Lake Michigan conference met in Chicago to discuss policy. EPA officials announced that the agency would issue interim discharge permits to the nuclear power plants operating on the lake, pending the establishment of guidelines for best practicable control technology for thermal discharges from steam electric power plants. In the meantime, the plants would be required to conform to the Lake Michigan temperature standards established by the states.[35]

The power companies also benefited from the reforms implemented by the AEC to streamline the licensing process. Pre-hearing conferences, regulations that imposed time limits on various stages of the hearing process and limited the kinds of challenges allowed, the use of separate rule-making hearings to examine generic technical questions—these and other changes significantly reduced the duration of licensing hearings. These reforms, along with a renewed determination by utility executives not to give in to what they saw as unreasonable demands, prevented BPI and other environmental groups in the Lake Michigan Basin from duplicating their earlier success in forcing nuclear power plants to install cooling towers.[36]

After much comment from industry, environmental groups, and other federal agencies on earlier drafts, the EPA issued its final effluent guidelines for steam electric power plants in October 1974. The regulations applied to all power plants that went on line after January 1, 1974, and to plants with a generating capacity greater than 500 megawatts that began operation after January 1, 1970. Under the guidelines, the plants would be required to implement closed-cycle cooling by July 1, 1983, at the latest. However, a provision of the 1972 act allowed power plants to utilize less stringent controls than those ultimately required by the EPA, if they could demonstrate that the promulgated requirements were "more stringent than necessary to assure the protection and propagation of a balanced, indigenous population of shellfish, fish, and wildlife." EPA officials planned to examine each plant on a case-by-case basis, but agency representatives made it clear they expected that most existing plants without closed-cycle cooling would be able to obtain such a variance. In the Lake Michigan Basin, all of the nuclear power plants that had managed to avoid the installation of cooling towers were allowed to continue operating without these systems.[37]

Implementation of the permit program created by the 1972 law also illustrated the continuing relevance of the state programs. The new law made July 1, 1977, the firm deadline for municipalities and industries to achieve secondary treatment and best practicable control technology, respectively. Nationwide, approximately 81 percent of the major industrial dischargers implemented BPT by the statutory deadline of July 1, 1977. In the Great Lakes Basin, the compliance rate was much lower, with only 58 percent of the significant industrial dischargers meeting the law's requirements in time. As expected, the majority of municipalities missed the law's deadline, with approximately 42 percent of all major municipal dischargers in the United States providing secondary treatment at that time. In the Great Lakes Basin, 41 percent of municipalities met the secondary treatment deadline.[38]

The new administration of Jimmy Carter assumed responsibility for administering the federal water pollution law only months before the July 1977 deadline came due. Environmentalists and other public interest organizations greeted the new administration with optimism and excitement, as the new president appointed more than sixty activists from public interest groups to government positions. As the July 1977 deadline loomed, Thomas Jorling, the recently appointed assistant administrator for Water and Hazardous Materials at EPA, promised "firm and fast enforcement action" against dischargers that failed to install the mandated treatment technology by July 1. Despite Jorling's tough rhetoric, however, state and federal authorities in the Great Lakes Basin did not take enforcement action against most of the municipalities that missed the July deadline, since the cities' lack of progress was attributed to the temporary impoundment of federal construction grant funds and administrative delays beyond the control of municipal officials.[39]

Determining the appropriate response to industries that failed to meet the July 1977 deadline was a more complex task. Environmentalists and others who hoped for an onslaught of lawsuits and fines directed against plants missing the compliance deadline were disappointed. When the deadline came, a number of large dischargers were still involved in adjudicatory hearings over the terms of their proposed permits, and the companies subsequently agreed to abatement schedules. In other cases, regulatory officials carefully examined the background and circumstances of each discharger. In some instances, the cases were referred to state and federal attorneys for legal action; but in most instances, mitigating circumstances and practical considerations led to deadline extensions without formal sanction.[40]

In spite of the intentions of legislators and the exhortations of environmentalists, regulatory officials—especially at the state level—continued to shy away from formal enforcement action and especially legal suits, except in the face of blatant defiance on the part of the discharger. The federal government continued to prod the states to take more aggressive action in enforcing water pollution control regulations. By the end of the decade, however, all of the states in the Great Lakes Basin had assumed authority over the administration of the NPDES program. While the EPA possessed the formal authority to review individual permits, take enforcement action, or even revoke state authority over the program, political constraints and limited manpower prevented the rigorous use of these powers.

The state programs were certainly much more willing to employ formal legal action against dischargers than before the upheavals of the Public Interest Era, but state officials (in most cases) exhibited much more flexibility than their federal counterparts. Ned Williams, director of the Ohio EPA, complained that "U.S. EPA's approach to enforcement seems to be that you sue anybody that has missed the . . . deadline. If there are good reasons for missing the deadline, that doesn't seem to matter." Michigan possessed one of the most effective state programs in the country. But in late 1977 a special internal task force that was created to examine Michigan DNR's enforcement program criticized the department's "excessive reliance on voluntary compliance efforts." Investigators found that staff members were reluctant "to pursue formal enforcement action on the premise that such action suggests 'failure' on the part of the Department to achieve compliance through negotiations and voluntary cooperation."[41]

Hesitance to pursue formal enforcement action may also have reflected the new salience of economic concerns in the 1970s, although state and federal regulators downplayed this factor. American industry had been on the defensive during the late 1960s and early 1970s, scrambling to adjust to demanding new requirements in environmental protection and other areas addressed by the New Social Regulation. By the mid-1970s, business leaders began to regroup, taking advantage of an "energy crisis" and economic recession to argue that overzealous government regulation was an important contributing factor to the nation's economic ills.[42] Business leaders in the Great Lakes region and other parts of the country placed particular emphasis on the cost of air and water pollution control, which they argued drove up energy costs, created inflationary pressures, and diverted funds from the more productive capital investments needed to combat increasingly stiff foreign competition. The importance of rational cost-benefit analysis was another frequently repeated theme. Too often (industry representatives argued) the benefits derived from large pollution control expenditures simply did not justify the direct and indirect costs. Such arguments were not new, but they achieved greater influence with elected officials in the context of America's economic difficulties.[43]

At the same time that industry executives were stepping up their lobbying activity to slow the pace of pollution abatement, public interest in the environment went into steady decline following the media hoopla of the first Earth Day and the major legislative victories of the early 1970s. Large segments of the population continued to support environmental protection efforts, but polling data from the period indicates a decline in the priority given to the environment relative to other problems. Murray Stein, now a special assistant in the EPA Water Office, summed up the prevailing mood in 1975 in his usual blunt fashion: "Earth Day 1975 is just barely alive. Energy has taken its place. The concern for the environment is [perceived as] a luxury."[44]

At the end of 1977, the lobbying of industry groups paid off when Congress enacted amendments to the federal law that relaxed compliance timetables and made other concessions to concerns about the economic impact of pollution abatement. The Clean Water Act of 1977, as it was known, authorized BPT deadline extensions up to April 1, 1979, for industrial dischargers who satisfied specific qualifications indicating a good faith effort to meet the original deadline. Under the amendments, industrial dischargers now faced regulations for three categories of pollutants—toxic, conventional, and nonconventional. The new law retained the strict BAT requirements and original July 1983 compliance deadline for toxic pollutants. In a concession to industry, dischargers of conventional pollutants such as suspended solids would be required to meet a new standard of treatment—"best conventional pollutant control technology"—that was to be at least as stringent as BPT, but not more stringent than BAT. Moreover, the compliance deadline for this class was extended to July 1984. Pollutants classified in the nonconventional category would have to meet BAT requirements by July 1987 at the latest.[45]

The 1977 amendments also provided that individual municipalities could be granted a deadline extension up to July 1, 1983, to achieve secondary treatment. To reach that goal, Congress authorized another $25.5 billion for municipal treatment plant construction grants through 1982. The amendments retained the 1983 compliance deadline for application of "best practicable waste treatment technology" by municipalities, even though EPA officials' most optimistic forecast was that general compliance with secondary treatment would not be attained until 1993. The Clean Water Act retained the no-discharge goal for 1985, but other sections of the act made clear that this objective had been effectively abandoned.[46]

Improvement in the Great Lakes

In 1976 EPA chief Russell Train referred to the recent progress in cleaning up the Great Lakes as "one of the greatest success stories in American history."[47] That a responsible public official could make such a statement would have seemed highly implausible, to say the least, just a few years earlier when Lake Erie was routinely referred to as a "dead sea" and the Great Lakes were held up as a national symbol of environmental disaster. But by the mid-1970s, many observers were expressing astonishment at the noticeable improvement in water quality in many parts of the Great Lakes Basin. Despite continuing problems at the city of Detroit sewage treatment system, the reappearance of game fish in the Detroit River and the dramatic improvement in the river's appearance—one middle-aged fisherman observed that for the first time in his life the water was actually blue—exemplified the turnaround. According to government data, the total daily volume of pollutants discharged to the Detroit River declined by approximately 82 percent between 1963 and the end of 1975. In Ohio, state authorities reported a steady decline in algae accumulation and bacteria levels at Lake Erie beaches. And in the summer of 1978, the Michigan DNR announced that major water quality improvements at Sterling State Park Beach on Lake Erie would allow the beach to open for swimming for the first time since 1961, when the closing of the beach led to the first federal enforcement conference in the Great Lakes region.[48]

Industrial plants in the Great Lakes Basin had long been the target of heated criticism for their failure to place greater controls on pollutant discharges and their slow compliance with government abatement orders. But in some of the most developed areas of the basin, the industrial cleanup initiated under the federal enforcement conferences and continued under the NPDES program was beginning to show results. Between 1972 and 1974, paper mills on the Fox River in Wisconsin reduced their discharge of suspended solids by over 50 percent and their BOD effluent content by almost 30 percent. At the end of 1978, 76 percent of the industrial plants in the American portion of the Basin were in compliance with BPT requirements, up from 58 percent in July 1977. Municipalities also made important strides forward. By the end of 1978, the percentage of the American population in the basin classified as receiving adequate treatment for their waste increased from 41 to 64 percent.[49]

Improvement in the efficiency of sewage treatment plants was especially important in controlling lake eutrophication. Progress on this front was slow but steady. By the end of the 1980s, additional legal restrictions on detergent phosphate content and improvements in phosphorus removal at municipalities had succeeded in reducing phosphorus levels below IJC objectives in Lakes Superior, Huron, and Michigan, and close to target values in Lakes Erie and Ontario.[50]

Regulatory authorities also made significant progress in reducing levels of toxic pollutants in the lakes, but growing scientific knowledge about their effects undermined any feeling of optimism among technical experts. In the first half of the 1970s, the IJC reported that state and federal usage and discharge bans had resulted in declining levels of DDT, dieldrin, and mercury in Great Lakes fish. However, PCBs and other toxins—such as the pesticide Mirex, which showed up in high levels in Lake Ontario fish—proved persistent in the face of control efforts. In 1977, the IJC stated that "the control and monitoring of toxic substances within the Great Lakes ecosystem is the most urgent problem facing the Governments under the present Water Quality Agreement."[51] The following year, the United States and Canada entered into a second Great Lakes Water Quality Agreement. More effective control of toxic pollutants was a central goal of the new agreement.

The improvement in Great Lakes water quality that first became evident in the 1970s has continued through the present day. The author of a recent article on pollution control efforts in the Great Lakes noted "that a region marred by black air and murky water during the 1960s and 1970s has become visibly cleaner. Beaches in the region no longer are regularly soiled by oily chemical residues washed up from factories around Chicago, Detroit and Gary, Ind." Much of this progress has resulted from significant declines in waste discharge from factories and municipal sewage treatment plants. Concentrations of DDT and other toxic compounds found in aquatic life and local wildlife have continued to decline.[52]

But serious problems continue. Some beaches in the region remain permanently closed. Combined sanitary and storm-water sewers in Detroit, Cleveland, and other lakefront cities still discharge raw sewage into local waters after modest rainfall. In industrial areas, bottom sediments are densely packed with toxic compounds, thus limiting aquatic life and threatening other water uses. Most important, whereas the levels of toxic contaminants have dropped, many scientists and public health officials are concerned that concentrations remain high enough to pose a long-term threat to all forms of life in the region. For this reason, the IJC has been pushing for a policy of "virtual elimination." Instead of trying to minimize toxic concentrations to "acceptable" levels, this policy would seek to ban the use or prevent any release of persistent toxic substances.[53]

The continuing efforts to deal with the ubiquity of toxic compounds have generated growing awareness about the need to link the problem of Great Lakes water quality to related problems in the broader ecosystem. Pollution from urban and especially agricultural runoff has become even more important in light of the progress made in controlling point sources of pollution. In spite of the Rust Belt image of the Great Lakes region, much of the land is devoted to crop production. Although the use of DDT and some other pesticides has been prohibited in the region, other types of pesticides continue to be applied in large volumes. Moreover, fertilizer use has increased significantly in the Great Lakes Basin in recent decades, as farmers attempt to secure greater crop yields per acre. Unfortunately, the regulatory methods used to limit point-source pollution are difficult to transfer to the farm sector, where voluntary participation or positive incentives (both technical and financial) have long been the norm in agricultural programs.[54]

In recent years, scientists have become more aware of the manner in which groundwater in the Great Lakes Basin eventually makes its way into surface waters, although the extent of this transfer depends on the geological characteristics of the different parts of the basin. This link is important for surface water quality in the Great Lakes Basin because of the numerous hazardous waste sites in the region that threaten to contaminate groundwater, as well as the fertilizers and pesticides that eventually end up in groundwater. Atmospheric deposition of toxic compounds is an even more frustrating problem. There is evidence that some contaminants deposited in the Great Lakes via air originate from as far away as Central and South America.[55]

Conclusion

In recent years, government regulation, particulary environmental protection, has come under increasing attack for being inefficient and ineffective. In the area of environmental policy, critics have charged Congress and the federal EPA with designing a regulatory system that emphasizes process over outcomes, gives priority to the wrong problems, and imposes rigid and inflexible rules on state programs and private industry.[56] The record of progess has indeed been mixed, although critics of the system tend to stress the most glaring example of failure. In the Great Lakes Basin, increasingly detailed federal laws, the greater willingness of government officials to take formal enforcement action against polluters, and the growth of knowledge about the Great Lakes ecosystem and the problems it faced eventually brought substantial improvement in water quality, although serious problems remain.

One of the major goals of this study has been to contribute to a greater understanding of the historical processes behind the major shift in regulatory policy—and indeed public policy in general—that took place during the Public Interest Era of the late 1960s and early 1970s. The New Social Regulation of the Public Interest Era addressed real problems. Environmental degradation in the Great Lakes region was quite visible in the early 1960s, but employment discrimination and inadequate workplace safety were just as obvious to those who encountered these problems. Of course, these and other problems addressed by the New Social Regulation had existed for some time, although certain forms of environmental pollution worsened in the post-1945 period. The critical point is that politicians and citizen reformers turned to issues such as environmental protection and product safety in the 1960s because these issues were closely tied to the search for a better quality of life that grew out of the affluent consumer society of postwar America.[57] For liberal politicians such as Lyndon Johnson and Gaylord Nelson, new laws designed to protect America's air and water, prevent discrimination, and improve access to medical care were all part of the quest to build a truly Great Society in the United States.

Affluence and a growing consumer orientation among large segments of the American population may have accounted for the broad social goals of these reforms, but the former trends do not explain why reformers tried to change the way policy was formulated and implemented in the United States. An important aspect of the Public Interest Era was the reformers' distrust of all authority, including the government, and their desire to open the administrative process to non-producer groups formerly excluded from these proceedings.[58] Developments in the Great Lakes Basin during this period help explain why these attitudes became so prevalent among reformers and demonstrate the dramatic impact these values had on public policy when they became enshrined in the regulatory system.

In the Great Lakes region after 1945, the state authorities administered a regulatory program that relied on voluntarism and informal cooperation and sought to balance the need for pollution control with other social and economic considerations. Clean water advocates, however, charged that state authorities were acting in the interests of major corporate polluters, not the public. During the early 1960s, federal officials pushed for higher levels of water quality across the nation and took steps to enhance their regulatory role. In the urban areas along the Great Lakes, local clean water advocates, the media, and some vocal members of

Congress called on the federal government to intervene in the region through the mechanism of the enforcement conference. Aided by new scientific revelations about the declining status of local water quality, critics of the state programs finally succeeded in having federal enforcement conferences convened to deal with pollution in some of the most polluted areas of the basin. From the end of 1965 through 1968, federal officials used the enforcement conferences and the new water quality standards program created by federal law to promote a general upgrade in waste treatment requirements in the Great Lakes Basin that departed from the reasonable use doctrine.

In post-1945 America, activists, newspaper editors, liberal federal legislators, and other parties seeking reforms of various kinds increasingly looked to the federal government as the vehicle for positive change. But this push for federal involvement was often followed by disillusionment with the effectiveness of federal programs. In the Great Lakes region, the federal enforcement conferences, initially seen as a turning point in efforts to halt the decline of water quality, became the focal point of dissatisfaction with government regulation, as major dischargers failed to meet compliance deadlines and local waters failed to show signs of improvement.

The system of cooperative pragmatism came apart during the Public Interest Era. State and federal attorneys began to intervene in the water pollution control process, seeking to circumvent with independent lawsuits the slow workings of the regulatory system. Concurrently, elected officials responded to public criticism by creating new regulatory institutions like the federal EPA that were often headed by young lawyers determined to establish their credentials with the growing environmental movement. These institutional reforms, however, did not completely assuage reformers' concerns about the "capture" of regulatory agencies by economic interest groups. Thus, the authors of the 1972 amendments to the Federal Water Pollution Control Act sought to limit agency discretion by specifying standards, mandating specific compliance deadlines, and creating opportunities for public interest groups to intervene in the regulatory process. This was the paradox of the Public Interest Movement: reformers wanted strong government to check private interests, but activists deeply distrusted the ability of the state to act in the broad public interest. A similar dynamic was also at work in the other areas of regulation that were restructured during this period. The long-term effects of this dichotomy on American society have been significant.

In the field of water pollution control, despite the reforms of the Public Interest Era, the principles of cooperative pragmatism did not fade away completely. Federal and state officials soon found themselves negotiating with industry representatives over effluent guidelines and individual permit requirements. And in spite of the federal primacy spelled out in the new law, state assumption of permit program administration ensured that economic concerns and other local needs would receive greater consideration than would have been the case with continued direct federal administration. This situation should come as no surprise. Most Americans have come to believe in the need for strong efforts to protect the environment. But in a liberal-capitalist system such as ours concerns about economic growth and the related value of individual economic security will always place certain limits on the ability and willingness of regulatory officials to impose costs on private economic interests, except in unusual cases of direct and unambiguous threats to public health.[59] Moreover, as much as some activists may prefer federal officials to take a large role in the administration of national laws, the small size of the federal bureaucracy relative to the laws and regulations it must oversee means that federal officials have no choice but to rely on their state counterparts to carry out the implementation of these laws.

Notes

Introduction

1. Events in Canada are obviously a vital aspect of the history of the Great Lakes. However, since I am most concerned with the development of environmental policy in the United States, I have chosen to deal with events in Canada only when they directly influenced policy making in the United States.

2. For an interpretive overview of changes in policy making during the 1960s, see Hugh Heclo, "The Sixties' False Dawn: Awakenings, Movements, and Postmodern Policymaking," *Journal of Policy History* 8, no. 1 (1996): 34–63. Like most scholars—and people in general—Heclo applies the term "1960s" in a loose manner that encompasses the early 1970s.

3. U.S. Bureau of the Census, *Census of Manufactures, 1963* (Washington, D.C.: GPO, 1966), vol. 1, pp. 10, 39, 43. Chicago was the second-largest standard metropolitan statistical area in the United States in terms of value added by manufacture. Detroit ranked fourth, Cleveland sixth, Milwaukee thirteenth, and Buffalo fourteenth. The top fifty metropolitan areas included smaller cities on the Great Lakes such as Rochester (nineteenth) and Toledo (thirty-seventh). In the 1963 ranking of the leading manufacturing states, seven of the top eleven states came from the Great Lakes region. New York ranked first, Ohio third, Illinois fourth, Pennsylvania fifth, Michigan sixth, Indiana eighth, and Wisconsin eleventh. Minnesota, the remaining Great Lakes state, occupied twentieth place on the list (ibid., p. 39).

4. For a concise overview of government-business relations in America, see the chapters on political economy in Mansel G. Blackford and K. Austin Kerr, *Business Enterprise in American History,* 2d ed. (Boston: Houghton Mifflin, 1990).

5. Some of the major works employing a broker state analysis are Theodore J. Lowi, *The End of Liberalism: The Second Republic of the United States,* 2d ed. (New York: W. W. Norton, 1979); Grant McConnel, *Private Power and American Democracy* (New

York: Alfred A. Knopf, 1967); and Otis L. Graham Jr., *Toward a Planned Society: From Roosevelt to Nixon* (New York: Oxford University Press, 1976). For a succinct description of American corporatism, see Michael J. Hogan, "Corporatism," in "A Round Table: Explaining the History of American Foreign Relations," *Journal of American History* 77, no. 1 (June 1990): 153–54. Robert Griffith, "Dwight D. Eisenhower and the Corporate Commonwealth," *American Historical Review* 87 (February 1982): 87–122, explains why corporatism was especially appealing to some American leaders in the 1950s. For an early elaboration of corporatism, see Ellis W. Hawley, "The Discovery and Study of a 'Corporate Liberalism'," *Business History Review* 52 (autumn 1978): 309–20. I realize that there are significant differences among these interpretations, but for the purposes of this study it is their common elements that are important.

6. George Hoberg, *Pluralism by Design: Environmental Policy and the American Regulatory State* (Westport, Conn.: Praeger, 1992), pp. 6–7 (7).

7. For the importance placed on "scientific expertise" in Progressive Era reform, see Samuel P. Hays, *Conservation and the Gospel of Efficiency: The Progressive Conservation Movement* (Cambridge, Mass.: Harvard University Press, 1959). William R. Childs, *Trucking and the Public Interest: The Emergence of Federal Regulation, 1914–1940* (Knoxville: University of Tennessee Press, 1985), also offers insight into the Progressive Era ideology of regulatory professionals (see especially pp. 87, 123).

8. The private members of the boards, it is important to note, were chosen because of their background and experience in business, municipal government, or other areas. These private members were representatives of interest groups in the broad sense that they supposedly shared the perspective and concerns of others in that group; they were not officers or official representatives from organizations such as the chamber of commerce or the farmers union.

9. Clarence Klassen, "Discussion," *Sewage and Industrial Wastes* 28 (February 1956): 218.

10. Robert L. Rabin, "Federal Regulation in Historical Perspective," *Stanford Law Review* 38 (1986): 1278–95; and David Vogel, "The 'New' Social Regulation in Historical and Comparative Perspective," in *Regulation in Perspective*, ed. Thomas K. McCraw (Cambridge, Mass.: Harvard University Press, 1981), pp. 155–86. On the Public Interest movement, see David Vogel, "The Public-Interest Movement and the American Reform Tradition," *Political Science Quarterly* 95, no. 4 (winter 1980–1981): 607–27; Robert Cameron Mitchell, "From Conservation to Environmental Movement: The Development of the Modern Environmental Lobbies," in *Government and Environmental Politics: Essays on Historical Developments since World War Two*, ed. Michael J. Lacey (Washington, D.C.: Woodrow Wilson Center Press, 1989), pp. 91–92; and Donald R. Brand, "Reformers of the 1960s and 1970s: Modern Anti-Federalists?" in *Remaking American Politics*, ed. Richard A. Harris and Sidney M. Milkis (Boulder, Colorado: Westview Press, 1989), pp. 37–38. The public interest groups focused especially on the corrupting power of big business in the American political system, echoing the preoccupation of many Progressive Era reformers. See Richard L. McCormick, "The Discovery that Business Corrupts Politics: A Reappraisal of the Origins of Progressivism," *American Historical Review* 86 (April 1981): 242–74.

11. Hoberg, *Pluralism by Design*, pp. 7–9; Heclo, "The Sixties' False Dawn," pp. 50–54; Hugh Davis Graham, "The Stunted Career of Policy History: A Critique and an Agenda," *Public Historian* 15, no. 2 (spring 1993): 22–26; and Marc Allen Eisner, "Discovering Patterns in Regulatory History: Continuity, Change, and Regulatory Regimes,"

Journal of Policy History 6, no. 2 (1994): 175–79.

12. Samuel P. Hays, *Beauty, Health, and Permanence: Environmental Politics in the United States, 1955–1985* (Cambridge, England: Cambridge University Press, 1987), pp. 36–39. On the environmental movement generally, see Kirkpatrick Sale, *The Green Revolution: The American Environmental Movement, 1962–1992* (New York: Hill and Wang, 1993); and Robert Gottlieb, *Forcing the Spring: The Transformation of the American Environmental Movement* (Washington, D.C.: Island Press, 1993). Jim O'Brien, "Environmentalism as a Mass Movement: Historical Notes," *Radical America* 17, no. 2–3 (March–June 1983): 7–27, is an insightful study.

13. For the tradition of the "radical amateur" in American environmentalism, see Stephen Fox, *John Muir and His Legacy: The American Conservation Movement* (Boston: Little, Brown, 1981).

14. According to James Q. Wilson, "entrepreneurial politics" is usually behind a policy that "will confer general (though perhaps small) benefits at a cost to be borne chiefly by a small segment of society." This equation holds true for many environmental laws and policies, although in the long run many of the costs involved are passed on to consumers. Wilson explains further that because "the incentive to organize is strong for opponents of the policy but weak for the beneficiaries, and since the political system provides many points at which opposition can be registered," skillful entrepreneurs must push these policies through to adoption by linking proposals to perceived crises and associating legislation with widely shared values such as personal safety. See James Q. Wilson, "The Politics of Regulation," in *The Politics of Regulation,* ed. Wilson (New York: Basic Books, 1980), p. 370.

15. Ibid., pp. 386–87.

16. Magali Sarfatti Larson observed that elected officials usually grant technical experts the most authority and autonomy in areas where there is the least conflict over policy. In his study of the engineers who staffed the Bureau of Public Roads, Bruce Seely found that the influence and power of the bureau declined after the number of interest groups making up the "highway community" expanded and the former consensus about means and goals fell to pieces. See Magali Sarfatti Larson, "The Production of Expertise and the Constitution of Expert Power," in *The Authority of Experts: Studies in History and Theory,* ed. Thomas L. Haskell (Bloomington: Indiana University Press, 1984), pp. 60–68; and Bruce E. Seely, *Building the American Highway System: Engineers as Policy Makers* (Philadelphia: Temple University Press, 1987).

17. William D. Ruckelshaus, *EPA Oral History Interview No. 1,* January 1993 (hereafter Ruckelshaus *Oral History Interview*), p. 9.

18. Joseph F. Zimmerman, *Federal Preemption: The Silent Revolution* (Ames: Iowa State University Press, 1991). For a broader discussion of changes in American federalism during the 1960s, see Martha Derthick, "Crossing Thresholds: Federalism in the 1960s," *Journal of Policy History* 8, no. 1 (1996): 64–80.

19. Hays, *Beauty, Health, and Permanence,* ch. 6; and Christopher Schroeder, "The Evolution of Federal Regulation of Toxic Substances," in Lacey, *Government and Environmental Politics,* pp. 275–77. Rachel Carson, *Silent Spring* (Boston: Houghton Mifflin, 1962).

20. The literature on risk assessment and environmental regulation is very large. For a succinct overview of this issue, see Walter Rosenbaum, *Environmental Politics and Policy,* 3d ed. (Washington, D.C.: Congressional Quarterly Press, 1995), ch. 5.

Chapter 1: "A Matter of Reasoned Cooperation"

1. Martin V. Melosi, "Environmental Crisis in the City: The Relationship between Industrialization and Urban Pollution," in *Pollution and Reform in American Cities, 1870–1930,* ed. Melosi (Austin: University of Texas Press, 1980), pp. 3–22. For a discussion of early stresses on the Great Lakes environment created by European settlement, see Thomas Ashworth, *The Late, Great Lakes: An Environmental History* (Detroit: Wayne State University Press, 1986).

2. *U.S. Eleventh Census,* vol. 1, p. xvii. From 1880 to 1890 Cleveland's population went from 160,000 to 261,000; Detroit's from 116,000 to 206,000; and Milwaukee's went from 116,000 to 204,000 (all figures rounded to the nearest thousand). In 1890, Chicago ranked as the second-largest city in America, Cleveland ranked tenth, Detroit fifteenth, and Milwaukee sixteenth.

3. Joel A. Tarr, "Industrial Wastes and Public Health: Some Historical Notes (Part 1, 1876–1932)," *American Journal of Public Health* 75, no. 9 (September 1985): 1059–60.

4. *First Annual Report of the State Board of Health, 1886, Ohio Executive Documents,* as excerpted in *An Ohio Reader: Reconstruction to the Present,* ed. Thomas H. Smith (Grand Rapids, Mich.: Eerdmans, 1975), p. 104. The report also described (p. 102) how waste discharges from Cleveland's oil-refining industry forced the city some years earlier to move the intake point for its water supply from a point a few hundred feet from the shore to a new location one mile from the shore.

5. Joel A. Tarr, James McCurley, and Terry F. Yosie, "The Development and Impact of Urban Wastewater Technology: Changing Concepts of Water Quality Control, 1850–1930," in Melosi, *Pollution and Reform,* pp. 59–70.

6. Ibid., p. 70; and Joel A. Tarr et al., *Retrospective Assessment of Wastewater Technology in the United States, 1800–1972: A Report to the National Science Foundation* (Pittsburgh, Pa.: Carnegie-Mellon University, 1977), ch. 3, pp. 1–2 (chapters are numbered separately in the report). Flow rate, temperature, chemical characteristics, and other factors also determine a stream's capacity to assimilate organic wastes. If the dissolved oxygen is exhausted, septic (anaerobic) conditions result. Bacteria will still decompose organic wastes under these conditions, but much more slowly, and the water will become putrid. (Sanitary professionals used the word "stream" as a generic term encompassing all bodies of surface water that received waste discharges. I will also follow this practice.)

7. Louis P. Cain, *Sanitation Strategy for a Lakefront Metropolis: The Case of Chicago* (DeKalb: Northern Illinois University Press, 1978), pp. xi–xii, 130; and Stuart Galishoff, "Triumph and Failure: The American Response to the Urban Water Supply Problem, 1860–1923," in Melosi, *Pollution and Reform,* p. 48.

8. James Joseph Flannery, "Water Pollution Control: Development of State and National Policy" (Ph.D. diss., University of Wisconsin, Madison, 1956), pp. 30–31; and Tarr et al., *Retrospective Assessment,* ch.4, pp. 6–9, 15.

9. Flannery, "Water Pollution Control," pp. 30–31.

10. Galishoff, "Triumph and Failure," pp. 44–53.

11. Ibid., pp. 53–54; and Tarr, McCurley, and Yosie, "Development and Impact," pp. 72–73.

12. John Duffy, *The Sanitarians: A History of American Public Health* (Urbana: University of Illinois Press, 1990), p. 129; and Tarr et al., *Retrospective Assessment,* ch. 4, pp. 8–9, 24–28. See also Joel A. Tarr, Terry Yosie, and James McCurley III, "Disputes

over Water Quality Policy: Professional Cultures in Conflict, 1900–1917," *American Journal of Public Health* 70, no. 4 (April 1980): 427–35.

13. Tarr, Yosie, and McCurley, "Disputes over Water Quality Policy," p. 429.

14. Ibid., pp. 429–30 (430).

15. Ibid., pp. 431–33.

16. Nandor Alexander Fred Dreisziger, "The International Joint Commission of the United States and Canada, 1895–1920: A Study in Canadian-American Relations" (Ph.D. diss., University of Toronto, 1974), pp. 288–90.

17. Cain, *Sanitation Strategy,* p. 127.

18. Dreisziger, "International Joint Commission," pp. 288–90.

19. For a general history of the IJC and American-Canadian environmental issues, see John E. Carroll, *Environmental Diplomacy: An Examination and a Prospective of Canadian-U.S. Transboundary Environmental Relations* (Ann Arbor: University of Michigan Press, 1983). Lake Michigan is entirely within the U.S. border and so was not included in the Boundary Waters Treaty of 1909 or subsequent studies.

20. International Joint Commission, *Final Report of the International Joint Commission on the Pollution of Boundary Waters Reference* (Washington, D.C.: GPO, 1918), pp. 18, 23, 27.

21. Ibid., pp. 36–37.

22. Dreisziger, "International Joint Commission," pp. 310–11.

23. Flannery, "Water Pollution Control," p. 31.

24. Tarr et al., *Retrospective Assessment,* ch. 7, pp. 4–5.

25. For the values of conservation professionals, see Hays, *Conservation and the Gospel of Efficiency.* In his dissertation on environmental politics in Wisconsin, Thomas R. Huffman argued that the professional training of Wisconsin pollution control officials led them to adopt a "production-centered" ethos similar to that of engineers and managers in natural resource–using industries. See Huffman, "Protectors of the Land and Water: The Political Culture of Conservation and the Rise of Environmentalism in Wisconsin, 1958–1970" (Ph.D. diss., University of Wisconsin, Madison, 1989), pp. 477–79.

26. Tarr et al., *Retrospective Assessment,* ch. 5, pp. 18–21.

27. O'Brien, "Environmentalism as a Mass Movement," 8–13. Historians usually refer to organizations from this period such as the Audubon Society and the Izaak Walton League as conservation groups. This term is potentially misleading, since it lumps these groups in with proponents of conservation such as Gifford Pinchot who emphasized the rational use of natural resources as economic commodities. In contrast, the citizen conservation organizations grew out of the John Muir wing of the conservation movement, which aimed at preserving the natural environment unspoiled, as something to be enjoyed and contemplated by future generations. For the formation and early activities of the Izaac Walton League, see Philip V. Scarpino, *Great River: An Environmental History of the Upper Mississippi, 1890–1950* (Columbia: University of Missouri Press, 1985).

28. Flannery, "Water Pollution Control," pp. 25–29; and Tarr, "Industrial Wastes and Public Health," pp. 1060–61.

29. Charles M. Henderson, "Statutory Developments in the Field of Water Pollution Control," in *Proceedings of the Sixth Industrial Waste Conference in Lafayette, Indiana, February 21–23, 1951,* Purdue University Engineering Extension Series No. 76 (West Lafayette: Purdue University, 1951) (hereafter *Industrial Waste Conference, February 1951*), pp. 45–47; and Flannery, "Water Pollution Control," pp. 33, 37–38, 41–42, 45.

30. Flannery, "Water Pollution Control," pp. 279–80; and Tarr et al., *Retrospective*

Assessment, ch. 7, pp. 11–12.

31. Tarr, McCurley, and Yosie, "Development and Impact," pp. 74–77.

32. Tarr, "Industrial Wastes and Public Health," p. 1061.

33. Ibid., pp. 1064–66. As Tarr notes, the emphasis on BOD as an indicator of industrial pollution neglected other important characteristics of industrial waste, especially with regard to inorganic pollutants.

34. Arthur E. Gorman, "Pollution of Lake Michigan: Survey of Sources of Pollution," *Civil Engineering* 3, no. 9 (September 1933): 519–21. Phenolic waste caused taste problems in the Chicago water supply and also absorbed the chlorine that was added to the water supply to protect it from contamination.

35. Langdon Pearse, "Chicago's Quest for Potable Water," *Water and Sewage Works* 102, no. 5 (May 1955): 190–91.

36. Tarr et al., *Retrospective Assessment,* ch. 7, pp. 16–19.

37. Flannery, "Water Pollution Control," p. 52. For the connection between postwar prosperity and environmentalism, see Samuel P. Hays, "From Conservation to Environment: Environmental Politics in the United States since World War Two," *Environmental Review* 6, no. 2 (fall 1982): 14–41. For the situation in state health departments, see Duffy, *The Sanitarians,* pp. 285–86.

38. Except for the Wisconsin State Committee on Water Pollution, each of these bodies was designated as either a board or a commission. To simplify matters, when these authorities are discussed in general terms, I will refer to them as boards.

39. Pennsylvania was unique among the Great Lakes states in that its Sanitary Water Board was technically a part of its health department, even though the board included representation from other state departments.

40. L. N. Rydland, "Michigan's Industrial Wastes Control Program," *Sewage and Industrial Wastes* 22 (December 1950): 1591; and Flannery, "Water Pollution Control," ch. 3.

41. International Joint Commission, *Report on the Pollution of Boundary Waters* (Washington, D.C., and Ottawa, 1951) (hereafter *IJC Report, 1951*), pp. 13–16. The other connecting channels were the St. Marys River (connecting Lake Superior and Lake Huron) and the Niagara River (connecting Lake Erie and Lake Ontario). A "reference" was a formal request by the governments of the United States and Canada to the IJC for an investigation and report on a particular issue or problem that fell under the purview of the Boundary Waters Treaty of 1909. My account of Detroit's role in initiating the study is drawn from the testimony of Nicholas Van Olds, Michigan assistant attorney general, in *Proceedings of the Joint Federal–State of Michigan Conference on Pollution of Navigable Waters of the Detroit River and Lake Erie, and Their Tributaries within the State of Michigan, First Session, Detroit, March 27–28, 1962* (Washington, D.C.: HEW, 1962) (hereafter *Detroit River Conference, March 1962*), pp. 352–53.

42. *IJC Report, 1951,* pp. 16–17.

43. Ibid., p. 17.

44. Ibid., pp. 168–69 (168).

45. Ibid., pp. 18–22.

46. Eugene W. Weber, "Activities of the International Joint Commission, United States and Canada," *Sewage and Industrial Wastes* 31 (January 1959): 74–77.

47. Flannery, "Water Pollution Control," pp. 146–70, 172–73.

48. Michael Joseph Donahue, *Institutional Arrangements for Great Lakes Management: Past Practices and Future Alternatives* (Ann Arbor: Michigan Sea Grant College Program, 1987), pp. 350–52.

49. Ibid., pp. 343–50. Continuing disagreement and litigation between Illinois and the other Great Lakes states over the amount of water diverted from Lake Michigan into the Chicago Sanitary and Ship Canal hampered interstate cooperation. The other states charged that Illinois's diversion caused declining water levels throughout the basin and interfered with navigation, hydroelectric power generation, and other water uses. See Beatrice Hort Holmes, *History of Federal Water Resources Programs and Policies, 1961–1970* (Washington, D.C.: DOA, 1979), pp. 66–69.

50. Tarr et al., *Retrospective Assessment,* ch. 7, pp. 19–23; and Flannery, "Water Pollution Control," pp. 388–90. This dissertation describes in great detail the legislative background of the 1948 act.

51. The major provisions of the 1956 act are explained in Curtiss M. Everts and Arve H. Dahl, "The Federal Water Pollution Control Act of 1956," *American Journal of Public Health* 47 (March 1957): 305–10. The following summary of the law's enforcement provisions is taken from Act of July 9, 1956, ch. 518, 70 *Statutes at Large,* pp. 504–5.

52. Act of July 9, 1956, ch. 518, 70 *Statutes at Large,* p. 498.

53. *Clean Waters for Ohio* 1, no. 1 (summer 1952): 6.

54. Henderson, "Statutory Developments," pp. 48–49. For a discussion of the increased importance of the administrative sphere in government regulation in modern America, see Samuel P. Hays, "Political Choice in Regulatory Administration," in *Regulation in Perspective,* ed. Thomas McCraw (Cambridge, Mass.: Harvard University Press, 1981).

55. Earl Finbar Murphy, *Water Purity: A Study in Legal Control of Natural Resources* (Madison: University of Wisconsin Press, 1961), p. 11; and Kenneth R. Jenkins, memo to Richard S. Green, August 8, 1961, PHS, RG 90 (hereafter PHS Records), A/N 66A-484, box 8, "Federal-State Relations" file, WNRC.

56. For Klassen, see *Who's Who in Engineering, 1964* (New York: Lewis Historical Publishing Company, 1964), pp. 1014–15, and "Mr. Pollution Control of Illinois: He's Optimistic," *Chicago Sun-Times,* January 19, 1970. For Poole, see *Who's Who in Engineering, 1959* (New York: Lewis Historical Publishing Company, 1959), p. 1960, and *Journal, WPCF* 37 (October 1965): 1453. For Oeming, see *Journal, WPCF* 37 (October 1965): 1455–56, and "Biographical Sketch" of Oeming, William Milliken Papers, box 756, "Environment Reorganization" file, Bentley Historical Library, University of Michigan.

57. Wilson, "Politics of Regulation," pp. 372–82. The two other basic types of regulatory employees, according to Wilson, were "careerists" and "politicians."

58. *Proceedings of the Conference in the Matter of Pollution of the Interstate Waters of the Grand Calumet River, Little Calumet River, Calumet River, Wolf Lake, Lake Michigan, and Their Tributaries, Chicago, March 2–9, 1965* (Washington, D.C.: HEW, 1965) (hereafter *Calumet Conference, March 1965*), pp. 627–30.

59. William J. Ronan, letter to U.S. Representative Charles Buckley, January 6, 1960, Nelson Rockefeller Papers (microfilm), reel 46, series 1959–1962, "Water Pollution, 1960–1961" file, New York State Archives. New York officials faced the greatest legal obstacles to enforcement action. Because of economic concerns, especially loss of industry, state legislators inserted a grandfather clause into the 1949 state water pollution law that made enforcement action against existing polluters almost impossible due to a complicated appeals process open to such dischargers. See the testimony of New York State health commissioner Hollis Ingraham in U.S. Congress, House Committee on Public Works, *Water Pollution Control: Hearings on Water Quality Act of 1965,* 89th

Cong., 1st sess., February 18, 19, 23, 1965 (hereafter *Water Quality Act Hearings, February 1965*), pp. 231–32, 255.

60. Murphy, *Water Purity,* pp. 105–6, 122.

61. Ruckelshaus *Oral History Interview,* pp. 5–7. The presence of special interests on most of the boards might also explain their cautious approach to regulation. In *Private Power and American Democracy,* published in 1967, Grant McConnel described how private interests used their representation on state government boards to shape policies that would coincide with their own narrow interests. In this case, however, the representatives of private interests were a minority on the pollution control boards, although they were certainly in a position to use their powers of persuasion on the other members of the board. Another point weighing against the importance of special interest representation is that the two states without private citizen membership on their boards—New York and Wisconsin—followed a cooperative approach that was very similar to those of the other states. Thus, the use of special interest representation on the boards should be viewed more as a manifestation of regulatory philosophy than as a determinant of regulatory action. See McConnel, *Private Power,* pp. 182–90.

62. Clarence Klassen, "Integrating a State Water Pollution Control Program with a Regional Water Resources Plan," *American Journal of Public Health* 43 (April 1953): 439–40 (439).

63. Clarence Klassen, "Water Quality Management—A National Necessity," in *Proceedings of the National Conference on Water Pollution in Washington, D.C., December 12–14, 1960* (Washington, D.C.: GPO, 1961) (hereafter *National Conference on Water Pollution, December 1960*), pp. 141–43 (142).

64. U.S. Congress, Senate Committee on Public Works, *Water Pollution Control: Hearings before the Special Subcommittee on Air and Water Pollution,* 88th Cong., 1st sess., June 17–20, 25–26, 1963 (hereafter *Water Pollution Control Hearings, June 1963*), pp. 416–17.

65. Flannery, "Water Pollution Control," p. 42; F. B. Milligan, "Standards for Treated Industrial Wastes," in *Industrial Waste Conference, February 1951,* pp. 114–15.

66. Testimony of John Vogt, director of the Division of Engineering, Michigan Department of Health, in *Detroit River Conference, March 1962,* p. 209. For examples of policy statements linking water use and assimilative capacity to setting treatment requirements, see the Indiana regulation reprinted in Jack E. McKee and Harold W. Wolf, eds., *Water Quality Criteria* (Sacramento, Calif.: Resources Agency of California, 1963), pp. 38–39; and John D. Porterfield, "Ohio Water Pollution Control Board—Its Policies and Administration," *Clean Waters for Ohio* 2, no. 1 (summer 1953): 7–8. On the importance of economic considerations, see Harold E. Babbit, "The Administration of Stream Pollution Prevention in Some States," in *Industrial Waste Conference, February 1951,* p. 247.

67. Dappert, memo to WPCB Members, October 28, 1954, New York Department of Environmental Conservation Records (hereafter EnCon Records), series A1120, box 5, "Erie-Niagara Classification Hearing" file, New York State Archives. Dappert discounted a conservationist's arguments against a Class E rating as lacking the "support of any basic facts." See also Roscoe C. Martin, *Water for New York: A Study in State Administration of Water Resources* (Syracuse: Syracuse University Press, 1960), ch. 7.

68. Dappert, memos to WPCB Members, February 16 and October 28, 1954, EnCon Records, series A1120, box 5, "Erie-Niagara Classification Hearing" file.

69. For the formation of the Cuyahoga River Basin Water Quality Committee, see Charles Lounsbury, plant manager, DuPont, letter to E. J. Riley, plant manager, Ferro

Chemical Corporation, May 3, 1965, and attachment, Ohio Water Pollution Control Board Records (hereafter Ohio WPCB Records), series 1800, box 10, file 2, Ohio State Archives. At the time of the letter, Lounsbury was cochairman of the committee. For Richards's statement, see Cuyahoga River Basin Water Quality Committee Minutes, May 26, 1965 (ibid.), pp. 2–3.

70. Quotation is from Flannery, "Water Pollution Control," p. 117. See also the comments of Milton Adams, MWRC executive secretary, in U.S. Congress, House Committee on Public Works, *Water Pollution Control Act: Hearings before the Subcommittee on Rivers and Harbors,* 84th Cong., 1st and 2d sess., July 20, 1955, and March 12–15, 1956, pp. 212–14; and Ralph E. Dwork, "Pollution 1955: A Report of Progress," *Clean Waters for Ohio* 3, no. 4 (spring 1955): 2, 6, 10.

Even a harsh critic of industry such as the Izaak Walton League noted the change. A representative from the league testified in the mid-1950s that industry "on the whole has in recent years, taken quite an enlightened view of its responsibilities," and that business firms were in some respects moving faster than municipalities in cleaning up their waste. Quoted in Flannery, "Water Pollution Control," p. 413.

71. See John A. Moekle, "Relation of Water Pollution Law Enforcement to Industrial Operation and Development," in *Proceedings of the Conference on Water Pollution Law Enforcement in Lansing, Michigan, November 22, 1965* (Lansing, Mich.: Attorney General, 1965). For almost identical comments, see the testimony of Jerome Wilkenfeld of Associated Industries of New York State, a state trade association, in House Committee on Government Operations, *Water Pollution Control and Abatement (Parts 1A and 1B— National Survey): Hearings before the Subcommittee on Natural Resources and Power,* 88th Cong., 1st sess., May–June 1963 (hereafter *Control and Abatement Hearings, May–June 1963*), pp. 1631–32.

72. See, for example, the testimony of the major steel, paper, petroleum, and chemical trade associations in House Committee on Government Operations, *Control and Abatement Hearings, May–June 1963.*

73. John Joseph Gargan, "The Politics of Water Pollution in New York State—The Development and Adoption of the 1965 Pure Waters Program" (Ph.D. diss., Syracuse University, 1968), pp. 137–40.

74. Indiana SPCB, *Annual Report, 1962–1963,* p. 4.

75. Anselmo Dappert, letter to Harry Hanson, November 9, 1955, EnCon Records, series A1120, box 4, "Lake Erie–Niagara River Abatement" file.

76. See the testimony of state officials at the congressional hearings on federal water pollution legislation in the 1940s and 1950s, especially that of Milton Adams, MWRC executive secretary. Adams was one of the leading water pollution control administrators in the country, and he was often chosen to represent the views of his counterparts in other states at these hearings.

77. Blucher Poole, "Roadblocks to Pollution Abatement," in *Proceedings of the Second State-Wide Water Supply Conference, in Chicago, August 27, 1958* (Illinois State Chamber of Commerce), p. 14. For a similar view, see the comments of Theodore Wisniewski, director of the Wisconsin Committee on Water Pollution, in *Proceedings of the State and Interstate Water Pollution Control Administrators in Joint Meeting with the Conference of State Sanitary Engineers in Washington, D.C., May 21, 1962* (HEW, PHS, Division of Water Supply and Pollution Control [DWSPC]) (hereafter *Sanitary Engineers Conference, May 1962*), p. 11.

78. A. J. Palladino, "Investigations—Pollution Abatement Measures beyond Primary

Treatment," in *Proceedings of the Fifteenth Industrial Waste Conference in Lafayette, Indiana, May 3–5, 1960,* (Purdue University Engineering Extension Series No. 106) (hereafter *Industrial Waste Conference, May 1960*) (West Lafayette: Purdue University, 1960), p. 351; and Milton P. Adams, "Water Resources as Viewed by a State Agency," *Journal, WPCF* 32 (May 1960): 523–24. The quote is from Palladino, "Investigations."

79. Adams, "Water Resources," 524.

80. Palladino, "Investigations."

81. MWRC Minutes, January 26, 1961, p. 5, and March 23, 1961, p. 7, Michigan DNR, Lansing. The minutes of the March meeting do not identify the "other interested parties."

82. MWRC Minutes, April 27, 1961, pp. 9–11.

83. Neil Staebler, letter to John Swainson, March 9, 1961, John Swainson Papers, box 14, "WRC: General" file, Bentley Historical Library, University of Michigan.

84. Joseph Buckley, letter to Swainson, July 7, 1961 (ibid.).

85. MWRC Minutes, June 22, 1961, p. 3.

86. MWRC Minutes, September 15, 1961.

87. Ibid.

88. MWRC Minutes, September 21, 1961, p. 2, and November 29, 1962, p. 7.

Chapter 2: "You Alone Have the Answer"

1. Ralph E. Dwork, "Pollution Control Is a Race," *Clean Waters for Ohio* 7, no. 1 (summer 1958): 8, and U.S. Congress, Senate Select Committee on National Water Resources, *Water Resources (Part 7),* 86th Cong., 1st sess., October 29, 1959 (hereafter *Water Resources Hearing, October 1959*), p. 1119. For a summary of treatment status in the Great Lakes Basin, as submitted by each of the Great Lakes states, see pp. 1114–23 of this hearing. Ohio authorities estimated that, together, municipalities and industries in the state spent more than $600 million on waste treatment between 1950 and 1960. See Ohio WPCB, *Annual Report for 1960,* p. 1.

2. *Water Resources Hearing, October 1959,* pp. 1114–23.

3. U.S. Bureau of the Census, *Census of Manufactures, 1963,* vol. 1 (Washington, D.C.: GPO, 1966), pp. 95–102.

4. The Detroit metropolitan area was a particularly striking example of this phenomenon. The population of the city of Detroit grew from 1,623,000 to 1,850,000 between 1940 and 1950, then dropped to 1,670,000 in 1960. In contrast, during that same twenty-year period, the population of Wayne County grew from 2,015,000 to 2,667,000. But it was the growth of the adjoining suburban counties of Macomb and Oakland that was truly impressive. The population of Macomb County grew from 108,000 to 406,000 between 1940 and 1960, whereas Oakland County's population increased from 254,000 to 690,000 over the same period. See Donald J. Bogue, *The Population of the United States: Historical Trends and Future Projections* (New York: Free Press, 1985), p. 120, and John L. Andriot, ed., *Population Abstract of the United States* (McLean, Va.: Andriot Associates, 1980), pp. 388–90.

5. Frank N. Egerton, "Missed Opportunities: U.S. Fishery Biologists and Productivity of Fish in Green Bay, Saginaw Bay, and Western Lake Erie," *Environmental Review* 13, no. 2 (summer 1989): 33–63. General data tracking Great Lakes water quality prior to the 1960s is fragmentary, but for summaries of evidence pointing to a long-term de-

cline, see Daniel A. Okun, "Managing the Great Lakes Water Resource," *Journal, WPCF* 41 (November 1969): 1859–60, and A. M. Beeton, "Eutrophication of the St. Lawrence Great Lakes," *Limnology and Oceanography* 10 (1965): 240–54.

6. Arve Dahl, "Water Pollution in the Great Lakes," address prepared for December 30, 1959, PHS Records, A/N 66A-484, box 11, "Speeches and Lectures" file, p. 3.

7. Ibid., pp. 3–5.

8. Martin V. Melosi, "Lyndon Johnson and Environmental Policy," in *The Johnson Years: vol. 2, Vietnam, the Environment, and Science,* ed. Robert A. Divine (Lawrence: University Press of Kansas, 1987), pp. 113–49; and Andrew J. Hurley, "Environmental and Social Change in Gary, Indiana, 1945–1980" (Ph.D. diss., Northwestern University, 1988), pp. 25–27. The statistics on the growth of the agency are drawn from "New Anti-Pollution Agency Faces Personnel Problems," *Engineering News-Record,* October 7, 1965, pp. 28–29. Construction grant funding made up the great bulk of the annual budget.

9. Leroy E. Burney cited from *National Conference on Water Pollution, December 1960,* p. 11; Gordon McCallum cited from "Filthy Water: Uncle Sam Steps Up Drive to Halt Pollution by Cities, Industries," *Wall Street Journal,* November 14, 1960; Mark Hollis, "Water Resources and Needs for Pollution Control," *Journal, WPCF* 32 (March 1960): 228. Hollis also called for new breakthroughs in waste treatment technology, since secondary treatment (or "complete" treatment, as it was sometimes referred to) only removed 90 percent of BOD and much less of other harmful pollutants. See also Mark D. Hollis and Gordon E. McCallum, "Dilution Is No Longer the Solution to Pollution," *Wastes Engineering* 30 (October 1959): 578.

10. Senate Committee on Public Works, *Water Pollution Control Hearings, June 1963,* p. 246. For statements citing concerns about driving industry out of state as a result of tough water pollution regulation, see testimony of Governor Nelson Rockefeller of New York, House Committee on Public Works, *Water Quality Act Hearings, February 1965,* pp. 232–34; and testimony of Michigan assistant state attorney general Nicholas Olds, U.S. Congress, House Committee on Public Works, *Water Pollution Control Act Amendments,* 88th Cong., 1st and 2d sess., December 1963 and February 1964, pp. 497–98.

11. Testimony of Nelson, Senate Committee on Public Works, *Water Pollution Control Hearings, June 1963,* p. 246; and testimony of Kennedy, U.S. Congress, Senate Committee on Public Works, *Water Pollution (Part 2): Field Hearings before the Special Subcommittee on Air and Water Pollution,* 89th Cong. 1st sess., June 17, 1965, pp. 792–93.

12. For representative samples of Dingell testimony, see House Committee on Government Operations, *Control and Abatement Hearings, May–June 1963,* p. 76; Senate Committee on Public Works, *Water Pollution Control Hearings, June 1963,* pp. 490–91, 501–2; and House Committee on Public Works, *Water Pollution Control Act Amendments, December 1963, February 1964,* pp. 55–57.

13. Kennedy, "Special Message to the Congress on Natural Resources, February 23, 1961," in U.S. President, *Public Papers of the Presidents of the United States: John F. Kennedy, 1961* (Washington, D.C.: GPO, 1962), pp. 116–17; Wilbur J. Cohen and Jerome N. Sonosky, "Federal Water Pollution Control Act Amendments of 1961," *Public Health Reports* 77 (February 1962): 107–8; and Melosi, "Lyndon Johnson and Environmental Policy," pp. 118–19.

14. Cohen and Sonosky, "Federal Water Pollution Control Act Amendments of 1961." Provisions to create a Federal Water Pollution Control Administration within

HEW were deleted from the water pollution legislation at the request of HEW secretary Abraham Ribicoff, who asked for more time to study the issue. See pp. 112–13.

15. Robert S. Hutchings, chief of Information Branch, letter to James Woodford, June 21, 1963, PHS Records, A/N 66A-484, box 21, "Public Awareness Program" file; "Water Pollution in the News: A Report, Summer 1963," ibid., "Reports" file; and Hutchings, letter to Robert Sansom, December 21, 1964, FWPCA, RG 382, WNRC (hereafter FWPCA Records), A/N 68A-1938, box 12, "Public Awareness Program" file. Testimony to the success of the public awareness program may also come from Peter Gammelgard of the American Petroleum Institute, who attacked the program in 1963 for exaggerating the nation's water pollution problem and minimizing the many accomplishments in this area. See House Committee on Government Operations, *Control and Abatement Hearings, May–June 1963*, p. 880.

16. *Pennsylvania Manual, 1955–1956* (Harrisburg, Commonwealth of Pennsylvania, 1956), p. 979; David R. Zwick and Marcy Benstock, *Water Wasteland: Ralph Nader's Study Group Report on Water Pollution* (New York: Grossman, 1971), pp. 57–59; and Cohen and Sonosky, "Federal Water Pollution Control Act Amendments of 1961," p. 113. According to Quigley, since other officials in HEW had more experience with well-established programs such as Social Security, Ribicoff tended to turn over responsibility for relatively new policy areas such as civil rights and pollution control to Quigley. See Quigley Oral History Interview, 1967 (John F. Kennedy Library), pp. 13–14.

17. *Environmental Health Letter,* January 15, 1962.

18. *Sanitary Engineers Conference, May 1962,* pp. 8–9; comments of Milton Adams in ibid., p. 30; Quigley Oral History Interview, p. 39.

19. The federal enforcement conferences are listed in Zwick and Benstock, *Water Wasteland,* Appendix A.

20. Testimony of Murray Stein, chief DWSPC enforcement officer, in Senate Committee on Public Works, *Water Pollution Control Hearings, June 1963,* pp. 79–80; and Holmes, *Programs and Policies,* p. 31, n. 15. See also Quigley Oral History Interview, pp. 25–26.

21. Murray Stein, "Enforcing the Federal Water Pollution Control Program," *Water and Sewage Works* 112 (October 1965): 354–57.

22. "Murray Stein: Top Pollution Cop," *Engineering News-Record,* October 7, 1965, pp. 54–55.

23. Ibid.; John Quarles, *Cleaning Up America: An Insider's View of the Environmental Protection Agency* (Boston: Houghton Mifflin, 1976), pp. 46–47; Zwick and Benstock, *Water Wasteland,* p. 122; and "Chairman Is Optimistic about Antipollution 'War'," *Cleveland Plain Dealer,* August 4, 1965.

24. *Calumet Conference, March 1965,* pp. 4–5.

25. *Proceedings of the Conference in the Matter of Pollution of the Navigable Waters of the Detroit River and Lake Erie, and Their Tributaries in the State of Michigan, Second Session, Detroit, June 15–18, 1965* (Washington, D.C.: FWPCA, 1965) (hereafter *Detroit River Conference, June 1965*), p. 28.

26. Senate Committee on Public Works, *Water Pollution Control Hearings, June 1963,* p. 50.

27. Holmes, *Programs and Policies,* pp. 68–69. Except for Indiana, the other Great Lakes states sought to require Chicago to end its diversion of Lake Michigan water or return the city's effluent to the lake to make up for the water lost through the diversion. The PHS study ultimately supported Chicago's position that discharge of the city's

treated effluent to the lake would have a significant negative impact on water quality. Chicago provided secondary treatment for its sewage, but the size of the population served by the system meant that the total waste volume would still be substantial. In December 1966, the Special Master appointed by the U.S. Supreme Court to consider the case ruled in Chicago's favor, and the status quo was maintained.

28. Testimony of Quigley in House Committee on Government Operations, *Control and Abatement Hearings, May–June 1963*, pp. 55–56; and Dean Coston, HEW, letter to Senator Charles Keating of New York, May 13, 1964, HEW, RG 235, WNRC (hereafter HEW Records), A/N 69A-1793, box 46, "Pollution, April–May 1964" file.

29. Lake Ontario Program Office, "Proposal for Program Implementation," August 16, 1965, FWPCA Records, A/N NC3-382-81-1, box 4, "Reports" file, pp. 4, 7.

30. Minutes of Technical Committee, December 1, 1961, pp. 4–5 and attachment, and September 7, 1962, PHS Records, A/N 66A-484, box 18, "Technical Committee to the GLIRB Project" file; DWSPC Staff Paper, April 15, 1964, FWPCA Records, A/N 68A-1938, box 27, "Water Pollution, Comprehensive Program" file, Attachment C, pp. 1–3; and Kenneth E. Biglene, PHS, memo to PHS regional health director, June 8, 1964, FWPCA Records, A/N 68A-1938, box 10, "Technical Committee to GRIRBP" file.

31. See, for example, the exchanges in Technical Committee, Minutes, September 7, 1962, PHS Records, A/N 66A-484, box 18, "Technical Committee to the GLIRB Project" file, pp. 2–6.

32. Blucher Poole, "Government Control of Industrial Wastes," in *Industrial Waste Conference, May 1960*, pp. 222–23 (222).

33. See Howard E. Moses, "Pennsylvania's Clean-Streams Program," *Journal, American Water Works Association* 42 (February 1950): 149; and L. N. Rydland, "Michigan's Industrial Wastes Control Program," *Sewage and Industrial Wastes* 22 (December 1950): 1591. See the comments of Frank Gregg, executive director of the Izaak Walton League, in *National Conference on Water Pollution, 1960*, pp. 94–99 (for AFL-CIO, p. 121).

34. Hays, "From Conservation to Environment"; O'Brien, "Environmentalism as a Mass Movement," pp. 15–17; and Hurley, "Change in Gary, Indiana," ch. 1.

35. O'Brien, "Environmentalism as a Mass Movement," p. 16. In Wisconsin, between 1945 and 1955, the number of fishing and hunting licenses distributed annually by the State Department of Conservation more than doubled, increasing from almost 700,000 to 1.7 million. See Huffman, "Protectors of the Land and Water," pp. 66–67.

36. This concept is an important theme of Hays, *Beauty, Health, and Permanence*, esp. pp. 36–39.

37. See *Proceedings of the Conference in the Matter of Pollution of Lake Erie and Its Tributaries, First Session, Cleveland, August 3–6, 1965* (Washington, D.C.: FWPCA, 1965) (hereafter *Lake Erie Conference, Cleveland, August 1965*), p. 107.

38. See, for example, the statement submitted by the Michigan United Conservation Clubs in *Detroit River Conference, March 1962*, pp. 945–48.

39. Robert Hutchings, memo to J. Stewart Hunter, September 23, 1963, PHS Records, A/N 66A-484, box 21, "Reports" file.

40. See Michael Schudson, *Discovering the News: A Social History of American Newspapers* (New York: Basic Books, 1978), ch. 5.

41. Curtiss M. Everts, "The Position of States in Water Pollution Control," *Journal, WPCF* 33 (February 1961): 161.

42. Drake is quoted in "The Press Agentry of Pollution," *Clean Waters for Ohio* 11,

no. 2 (summer 1962): 14–15.

43. *Detroit News,* May 12, 1962.

44. *Cleveland Plain Dealer,* April 3, 1960.

45. Ralph Dwork, memo to Governor DiSalle, March 23, 1960, Michael DiSalle Papers, box 43, "Dept. of Health, Director" file, Ohio State Archives.

46. Poole cited from Steering Committee, National Association of Administrators of State and Interstate Water Pollution Control Administrators, Minutes, February 6–7, 1963, PHS Records, A/N 66A-484, box 18, "Organizations, Committees, and Conferences—S" file. Oeming, "Water Quality Management: Administrative Aspects," *Journal, WPCF* 36 (September 1964): 30.

47. MWRC, *Quarterly Bulletin,* no. 7 (quarter ending March 31, 1960): 1.

48. State controller Ira Polley, memo to Governor Swainson, August 3, 1961, Swainson Papers, box 14, "Detroit River Pollution" file; and *Detroit River Conference, March 1962,* pp. 362–65.

49. These generalizations are based on letters submitted for the record at the Detroit River federal enforcement conference by individuals, civic associations, local conservation groups, and other parties from this area. See *Detroit River Conference, March 1962,* pp. 356–584. During 1961 Michigan experienced the greatest outbreak of hepatitis in its history, as did the United States as a whole. Monroe County was particularly hard hit, but the county health director denied any connection with polluted lake water, maintaining that all evidence pointed to person-to-person contact as the means for spreading the disease (ibid., pp. 759–75).

50. "Start Drive to Stop Lake Erie Pollution," *Detroit News,* September 14, 1961.

51. John Chascsa, letter to HEW Secretary Ribicoff, December 15, 1961, Swainson Papers, box 33, "Detroit River Pollution" file. The minutes of the second through sixth meetings of the Lake Erie Cleanup Committee are reprinted in *Detroit River Conference, March 1962,* pp. 599–631. These meetings were held once a month from September 1961 through February 1962. The respect accorded the committee's influence is evident in appearances by officials from the MWRC, the Michigan Department of Health, the PHS, the city of Detroit, state legislators, and some local industries.

52. *Detroit River Conference, March 1962,* pp. 498–511. Coliform bacteria were not in themselves harmful, but their presence in large numbers served as an indicator of fecal pollution and the possible presence of pathogenic organisms.

53. Hart, letter to Swainson, October 18, 1961, Swainson Papers, box 14, "WRC: General" file. On the similar urging by Representative Dingell, Senator Pat McNamara (Democrat) and downstream state legislators, see Milton Adams, memo to Water Resources Commission, November 29, 1961, Swainson Papers, box 14, "Detroit River Pollution" file, pp. 3–4.

54. Adams, letter to Swainson, November 2, 1961, and Boyd Benedict, governor's staff, memo to Swainson, November 7, 1961, Swainson Papers, box 14, "Detroit River Pollution" file.

55. Adams, letter to Water Resources Commission, November 14, 1961, and Benedict, memo to Swainson, November 15, 1961, ibid.

56. Swainson, letter to Ribicoff, December 5, 1961, George Romney Papers, box 336, "Great Lakes States Conference, May 1965" file, Bentley Historical Library, University of Michigan. After the conference had been scheduled, Michigan officials—joined by Gerald Remus of the Detroit sewer authority—tried to get Swainson to cancel the conference or at least postpone it for several years while Detroit implemented a plan to improve

municipal treatment in the Detroit area. See Benedict, memo to Swainson, January 31, 1962, Swainson Papers, box 33, "Detroit River Pollution" file.

57. "U.S. Agent Promises Action on Pollution," *Grand Rapids Press,* December 17, 1961.

58. *Detroit River Conference, March 1962,* pp. 63–64, 98–103.

59. Ibid., pp. 103–4, 5–7.

60. Ibid., pp. 424–27, 456–61. Remus also questioned how much gain secondary treatment would bring, when most of the Canadian cities discharging to these waters provided no treatment for their sewage.

61. See statement by Adams in *Sanitary Engineers Conference, May 1962,* p. 32; and letter, John Vogt to Karl Mason, January 1, 1963, EnCon Records, series A1118, box 18, "Water Resources: State and Interstate Pollution Control Administrators" file.

62. Office of Senator Hart press release, October 21, 1963, Romney Papers, box 26, "WRC: General" file; MWRC Minutes, October 31, 1963, p. 2; and "Pollution Control Praised," *Detroit News,* May 2, 1963.

63. For the general support given to federal enforcement conferences by local newspaper editors, see "Water Pollution in the News: A Report, Summer, 1963," PHS Records, 66A-484, box 21, "Reports" file, p. 3. Increasing media attention to water pollution control is discussed in Robert S. Hutchings, chief of DWSPC Information Branch, memo to J. Stewart Hunter, assistant to surgeon general for Information, June 24, 1963, ibid.

64. "U.S. Getting Tougher in Enforcing Cleanup of Polluted Waterways," *Wall Street Journal,* October 14, 1963; "Cleaner Water Drive Steps Up," *Business Week,* March 13, 1965, p. 78.

65. Oeming, letter to Governor Romney, August 12, 1963, Romney Papers, box 25, "Water Resources Commission" file. Once again, Michigan Senators Hart and Mc-Namara viewed the convening of a federal enforcement conference in Michigan in a favorable light. See Pat McNamara, letter to Oeming, October 8, 1963, and a Hart press release, October 21, 1963, Romney Papers, box 25, "Water Resources Commission" file. McNamara's reply to Oeming suggests that Oeming had contacted him about the possibility of having the conference postponed.

66. "Stream Improvement Praised in Wisconsin," *Pulp and Paper,* April 16, 1962, pp. 9–10.

67. *Proceedings of the Conference in the Matter of the Pollution of the Interstate Waters of the Menominee River and Its Tributaries, Menominee, Michigan, November 6–8, 1963* (Washington, D.C.: HEW, 1964), p. 778.

68. W. F. Carbine, statement in *Detroit River Conference, March 1962,* pp. 143–44; "The Furore about Lake Erie," *Clean Waters for Ohio* 11, no. 1 (spring 1962): 12–13. The earlier work of fishery biologists in the Great Lakes emphasized overfishing rather than pollution as the major cause of commercial fishing woes. See Egerton, "Missed Opportunities." Fish continued to be plentiful in Lake Erie, but their total population had shifted from a predominance of high-value species such as pike and white fish to low-value fish such as carp and smelt. The introduction of exotic fish species was another factor behind the industry's decline.

69. For an extended technical discussion of eutrophication, see IJC, *Pollution of Lake Erie, Lake Ontario, and the International Section of the St. Lawrence River* (Washington, D.C.: IJC, Canada, and the United States, 1970) (hereafter *IJC Report, 1970*), vol. 1, pp. 21–31.

70. "Furore about Lake Erie," 12–13.

71. Drake is quoted in "The Press Agentry of Pollution," *Clean Waters for Ohio* 11, no. 2 (summer 1962): 13. For the large algae deposits on Lake Erie beaches, see MWRC, *Quarterly Bulletin* no. 36 (quarter ending June 30, 1962): 5.

72. Cleveland's newspaper articles cited in John H. Puzenski, Cuyahoga County sanitary engineer, "A Report on Pollution Control Progress in Cuyahoga County," June 14, 1961, Ohio WPCB Records, series 1800, box 16, file 9, p. 12; editorial, "Pollution and Politics," *Cleveland Plain Dealer,* February 17, 1962.

73. Harold Titus, "The Fight to Save Lake Erie," *Field and Stream* (March 1965): 10–11.

74. H. W. Poston, DWSPC regional program director, memo to Gordon McCallum, July 14, 1964, FWPCA Records, A/N 68A-1938, box 18, "Lake Erie" file; *Cleveland Plain Dealer,* July 29, 1964; "Pollution Foe Defends Use of Billboard," *Cleveland Plain Dealer,* July 30, 1964. An outdoor advertising company donated twelve billboards to Blaushild's campaign, stretching along the Ohio lakeshore from Conneaut to Port Clinton.

75. These resolutions can be found in FWPCA Records, A/N 68A-1938, box 18, "Lake Erie" file. By November, an additional seventeen Ohio communities had passed the Blaushild resolution. See "Ohio's Troubled Waters," *News in Engineering* (November 1964): 7.

76. "Locher Asks for Delay in Anti-Pollution Action," *Cleveland Press,* August 1, 1964; and "Pollution Policy Defended," *Cleveland Plain Dealer,* September 26, 1964.

77. Northington, "Record of Meeting at City Hall, Cleveland, Ohio, on September 3, 1964," HEW Records, A/N 69A-1793, box 46, "Pollution, September–December 1964" file. Mayor Locher was not present at the meeting but later confided to Northington that he felt enactment of the resolution would make the mayor look a fool.

78. "Pollution Policy Defended," *Cleveland Plain Dealer,* September 26, 1964; "Call Better Sewers Key to Clean Lake," *Cleveland Press,* September 25, 1964; and Northington, "Meeting with Cleveland City Council Committee on Air and Water Pollution, September 25, 1964," FWPCA Records, A/N 68A-1938, box 18, "Lake Erie" file, p. 3.

79. Northington, "Meeting with Cleveland City Council Committee on Air and Water Pollution, September 25, 1964," p. 4. When Northington reported he had told the local parties that evidence of substantial progress and cooperation by the city of Cleveland and local industries could prevent the convening of an enforcement conference, he received a sharp rebuke from regional program director Wally Poston, who stated that only the secretary had the authority to make such decisions and that personnel in the field should make no assurances of this kind. See Poston, memo to W. Q. Kehr, director, Great Lakes–Illinois River Basin Project, October 6, 1964, ibid.

80. Vanik's letter of March 2, 1965, is reprinted in *Lake Erie Conference, Cleveland, August 1965,* pp. 25–26. The other letters are in HEW Records, A/N 69A-1793, box 46, "Pollution, January–March 1965" and "Pollution, April–June 1965" files.

81. For the expanding media coverage of Lake Erie, see James Barnhill, DWSPC deputy chief, memo to Quigley, March 31, 1965, FWPCA Records, A/N 68A-1938, box 18, "Lake Erie" file, p. 1. Lyndon B. Johnson, "Special Message to the Congress on Conservation and Restoration of Natural Beauty, February 8, 1965," in U.S. President, *Public Papers of the Presidents of the United States: Lyndon B. Johnson,* Book 1 (Washington, D.C.: GPO, 1966), p. 162.

82. "Critical Pollution Is Found in Lake," *Cleveland Press,* March 25, 1965.

83. Quigley, letter to Rhodes, April 22, 1965, Anthony J. Celebrezze Papers, series 2, box 17, folder 327, Western Reserve Historical Society, Cleveland.

84. "Proceedings of the Governors Conference on Great Lakes Pollution, in Cleveland, May 10, 1965," State Library, Columbus, Ohio.

85. Ibid., pp. 168–72.

86. Ibid., pp. 56–68.

87. Ibid., 68–69.

88. Rhodes, letter to Celebrezze, June 11, 1965, James Rhodes Papers, box 14, folder 6, Ohio State Archives. By this time, HEW officials believed that there was enough evidence of interstate pollution to justify the inclusion in the conference of all the states discharging to Lake Erie.

89. U.S. Congress, House Committee on Government Operations, *Water Pollution Control and Abatement (Part 3—Chicago Area and Lower Lake Michigan): Hearings before the Subcommittee on Natural Resources and Power,* 88th Cong., 1st sess., September 6, 1963, pp. 2086–91, 2103.

90. Ibid., pp. 2214–18. The major discharger in Illinois territory was the U.S. Steel South Works.

91. Articles from the *Chicago Daily Calumet* are reprinted in ibid., pp. 2345–51.

92. Yoder, letter to Kerner, May 6, 1964, Otto Kerner Papers, box 277 (only one folder in box), Illinois State Historical Society Archives, Springfield.

93. "Illinois, Indiana Map Fight on Lake Michigan Pollution," *Chicago Daily News,* September 10, 1964.

94. Quigley's recommendation for the convening of a conference cited staff reports that discharges into Lake Michigan from both Indiana and Illinois were interfering with public water supply, damaging commercial and sport fishing, and creating a visual nuisance. The existence of such conditions and the extent of interstate pollution had been obvious for some time. Quigley, memo to Celebrezze, December 15, 1964, EPA, RG 412, WNRC (hereafter EPA Records), A/N 74-22, box 47, "Lake Michigan (Calumet), Admin." file no. 1.

95. Bacon is quoted in "Tell Water Pollution Peril Here," *Chicago Tribune,* December 28, 1964.

96. Oeming, "State and Federal Cooperation in Water Quality Management," June 17, 1965 (typed manuscript, Michigan State Library, Lansing).

Chapter 3: Standards and Deadlines

1. *Report on Pollution of the Detroit River, Michigan Waters of Lake Erie, and Their Tributaries: Summary, Conclusions, and Recommendations, April 1965* (Washington, D.C.: HEW, 1965) and *Report on Pollution of Lake Erie and Its Tributaries, July 1965* (Washington, D.C.: HEW, 1965). The report on the Calumet area and southern Lake Michigan is reprinted in *Calumet Conference, March 1965,* vol. 1.

2. See reports cited in previous note. Obtaining useful data from industrial dischargers was a major headache for federal and (to a lesser extent) state officials. It was not until the passage of the 1972 amendments to the Federal Water Pollution Control Act that all major industries faced stringent legal requirements for providing regulators with comprehensive, up-to-date information on waste characteristics, effluent volume, and treatment efficiencies. See Zwick and Benstock, *Water Wasteland,* ch. 12.

3. Statistics are taken from *Report on Pollution of the Detroit River,* p. 13; "Catalyst for a Cleaner Lake," *Cleveland Plain Dealer,* August 3, 1965.

4. Federal representative at Calumet conference cited in "Pollution Fighters Close Ranks," *Engineering News-Record,* March 11, 1965, p. 23; *Lake Erie Conference, Cleveland, August 1965,* pp. 739–41 (Rhodes's statement pp. 17–20; Arnold quoted on p. 614).

5. *Proceedings of the Conference in the Matter of Pollution of Lake Erie and Its Tributaries, First Session, Buffalo, August 10–11, 1965* (Washington, D.C.: FWPCA, 1965) (hereafter *Lake Erie Conference, Buffalo, August 1965*), p. 137.

6. *Detroit River Conference, June 1965,* pp. 1506–9 (Tucker), 1564–69 (Black).

7. *Lake Erie Conference, Cleveland, August 1965,* pp. 1065–74 (1070).

8. For Remus's arguments, see *Detroit River Conference, June 1965,* pp. 1383–404; "Pollution Report Hit by Remus," *Detroit News,* May 8, 1965, and "Pollution War Costs to Soar, Remus Warns," *Detroit Free Press,* June 19, 1965. Final recommendations are from *Detroit River Conference, June 1965,* p. 1784.

9. "Water Pollution Victory," *Chicago American,* March 11, 1965 (quote); and "Water Confab Set the Stage," *Cleveland Plain Dealer,* August 7, 1965. H. W. Poston, "A Close Look at a Recent Federal Enforcement Conference," in *Proceedings of the Twentieth Industrial Waste Conference in Lafayette, Indiana, May 4–6, 1965,* Purdue University Engineering Extension Series No. 118 (West Lafayette: Purdue University, 1965) p. 3.

10. Eagle cited from Cuyahoga River Basin Water Quality Committee, Minutes, September 22, 1965, Ohio WPCB Records, series 1800, box 10, file 2, p. 3.

11. Oeming, "State and Federal Cooperation," pp. 2–3.

12. MWRC Minutes, June 24, 1965, p. 6; July 29, 1965, pp. 6–7; and August 25, 26, 1965, p. 7. "Rouge River Is Hopeless, Pollution Fighters Admit," *Detroit News,* August 27, 1965. For a graphic description of the Rouge River, see "We've Created a Water Wasteland," *Detroit News,* July 24, 1966.

13. Harlow quoted in "Trace Water Pollution to Kitchens," *Detroit News,* August 26, 1965. See also "Water Aides, Fifteen Industries Discuss Pollution Problems," *Detroit News,* October 3, 1965; and Harlow, memos to Murray Stein and Gordon McCallum, October 6 and November 17, 1965, EPA Records, A/N 74-22, box 47, "Detroit River, Admin." file no. 5.

14. "State Board Takes a Holiday, Puts Off Fighting Pollution," *Detroit News,* December 17, 1965; Herb DeJong, memo to Romney, December 16, 1965, Romney Papers, box 336, "Pollution of Lake Erie and Detroit River" file; and MWRC Minutes, January 6, 1966, pp. 2–3.

15. MWRC Minutes, March 30, 31, 1966, pp. 5–6; for Oeming's warning, see "Excuses Fail to Halt State in Drive to Clean Up River," *Detroit News,* March 31, 1966.

16. MWRC, *Quarterly Bulletin,* no. 52 (June 1966): 1. The one holdout was Scott Paper, which agreed to execute a stipulation with the MWRC in September to end its pollution of the Rouge River. The MWRC agreed to drop enforcement proceedings initiated against the company. See MWRC, *Quarterly Bulletin,* no. 53 (September 1966): 1.

17. Remus, letters to MWRC, March 28, 1966, and to Oeming, March 31, 1966, EPA Records, A/N 74-22, box 27, "Detroit River, Admin." file no. 5. Stein is quoted from U.S. Congress, Senate Committee on Public Works, *Water Pollution Control—1966: Hearings before the Subcommittee on Air and Water Pollution,* 89th Cong., 2d sess., April–May 1966, p. 435.

18. For a listing of committee members and alternates, see *Proceedings of the Conference in the Matter of Pollution of the Interstate Waters of the Grand Calumet River, Little*

Calumet River, Calumet River, Wolf Lake, Lake Michigan, and Their Tributaries, Chicago, Technical Session, January 4–5, 1966 (Washington, D.C.: FWPCA, 1966) (hereafter *Calumet Conference, January 1966*), pp. 107–8.

19. Ibid., pp. 142–54.

20. *Proceedings of the Conference in the Matter of Pollution of the Interstate Waters of the Grand Calumet River, Little Calumet River, Calumet River, Wolf Lake, Lake Michigan, and Their Tributaries, Chicago, Technical Session, February 2, 1966* (Washington, D.C.: FWPCA, 1966) (hereafter *Calumet Conference, February 1966*), pp. 54–55; and Indiana SPCB Minutes (Indiana Department of Environmental Management, Indianapolis), Book 29, January 12, 1966, pp. 107–8, and February 3, 1966, p. 139. See also "Area Water Picture Still Muddied," *Gary Post-Tribune*, February 23, 1966.

21. "Lake Michigan Accord Hailed," *Christian Science Monitor*, February 16, 1966.

22. *Proceedings, Third Meeting of the Conference in the Matter of Pollution of Lake Erie and Its Tributaries, Cleveland, June 22, 1966* (Washington, D.C.: FWPCA, 1966) (hereafter *Lake Erie Conference, June 1966*), pp. 578–606. The federal enforcement conference on the Detroit River and the Michigan waters of Lake Erie was absorbed into the five-state conference on Lake Erie, so follow-up on the recommendations for the Detroit River area took place within the latter conference.

23. *Proceedings, Third Session of the Conference in the Matter of Pollution of Lake Erie and Its Tributaries, Buffalo, March 22, 1967* (Washington, D.C.: FWPCA, 1967) (hereafter *Lake Erie Conference, March 1967*), pp. 482–84.

24. "New Anti-Pollution Agency Faces Personnel Problems," *Engineering News-Record*, October 7, 1965, pp. 28–29; and Zwick and Benstock, *Water Wasteland*, p. 58.

25. Udall, memo to President Johnson, September 2, 1965, and Joseph Califano, memo to President Johnson, September 8, 1965, White House Central Files (hereafter WHCF), HE 8-1, box 2, "Air Pollution" file, Lyndon Baines Johnson Presidential Library (hereafter Johnson Library). For the arguments against the transfer to the Department of Interior, see HEW secretary John Gardner, memo to Califano, December 27, 1965, and Lee White, memo to President Johnson, January 12, 1966, WHCF, EX FG 165-6-3, box 250, "Water Pollution Control Advisory Board" file, Johnson Library.

26. Melosi, "Lyndon Johnson and Environmental Policy," pp. 120–21; and Stewart Udall, *The Quiet Crisis* (New York: Holt, Rinehart, and Winston, 1963), pp. vii–viii. See also Peter Wild, *Pioneer Conservationists of Western America* (Missoula, Mont.: Mountain Press, 1979), ch. 15; and Barbara Le Unes, "Stewart Lee Udall," in *Encyclopedia of American Forest and Conservation History*, ed. Richard C. Davis (New York: Macmillan, 1983), pp. 665–66.

27. For Udall's emphasis on educating the public, see Henry B. Sirgo, "Water Policy Decision-Making and Implementation in the Johnson Administration," *Journal of Political Science* 12, no. 1–2 (1985): 54–57.

28. *Lake Erie Conference, June 1966*, pp. 7–8 (7). See also the comments by assistant secretary of the Interior Frank DiLuzio, in Frank DiLuzio, "Water Pollution Control—An American Must," *Journal, WPCF* 39 (January 1967): 2–3.

29. "LBJ Views Harbor and Sees a Sample of River Pollution," *Buffalo Evening News*, August 20, 1966; and "Text of the Remarks of the President at Niagara Square, Buffalo, New York," WHCF, Office Files of Ceil Bellinger, box 28, "Pollution" file, Johnson Library.

30. "The Brass Blasts at Pollution," *Chemical Week*, October 9, 1965, p. 25; and

"Report of the Task Force on Pollution Abatement," October 9, 1965, Bureau of Budget Records, RG 51 (hereafter BOB Records), series 61.1, 1965, "Water Pollution" file, pp. 19–27, Washington National Archives. Secretary Udall met with the task force several times during its deliberations.

31. "LBJ Escalates War on Pollution," *Engineering News-Record,* March 3, 1966, p. 11; "Big Spending on Waste Purification Pushed by Key Lawmakers, Bucking Administration," *Wall Street Journal,* June 23, 1966.

32. Holmes, *Programs and Policies,* pp. 186–93. For discussion of the legislative debate prior to the passage of the Water Quality Act, see Philip P. Micklin, "Water Quality: A Question of Standards," in *Congress and the Environment,* ed. Richard A. Cooley and Geoffrey Wandesforde-Smith (Seattle: University of Washington Press, 1970), pp. 130–47.

33. James L. Agee and Allan Hirsch, "Water Quality Standards—The Role They Will Play in Administering Water Pollution Control Programs," in *Proceedings of the Twenty-Second Industrial Waste Conference in Lafayette, Indiana, May 2–4, 1967,* Purdue University Extension Series No. 129 (West Lafayette: Purdue University, 1967), pp. 12–15. Determining the industry equivalent of secondary treatment was a relatively straightforward matter when dealing with paper mills and other industrial dischargers that generated organic waste. But in cases where the impact of pollutants could not be measured in terms of BOD, this became a much more arbitrary process. Guideline No. 8, which dealt with this issue, called on dischargers to apply the "best practicable treatment" to their wastes unless they could demonstrate that a lesser degree of treatment would protect present and proposed water uses. Basing treatment requirements on technological feasibility later became federal law (see epilogue).

34. "New U.S. Agency and New Policy to Enter Fight against Water Pollution," *New York Times,* December 21, 1965. See also DiLuzio, "Water Pollution Control—An American Must," p. 4.

35. This summary of state procedures is based on relevant documents contained in each of the state archives consulted for this study.

36. Ibid.

37. Transcript of Proceedings, MWRC Hearing, April 27, 1967, in Detroit, Romney Papers, box 337, "Water Quality Standards for St. Clair River, etc." file; and Transcript of Proceedings, Indiana SPCB Hearing, April 10, 1967, in Gary, Indiana, Indiana SPCB Records (hereafter Indiana SPCB Records), River Basin files, box R2258, "Lake Michigan Basin" file, Indiana State Archives. The Indiana SPCB established interstate water quality standards for the ship canal because the canal is connected to the Grand Calumet River, which spans the Illinois-Indiana border.

38. "Public Hearing on Water Quality Standards for the Rocky, Cuyahoga, Chagrin, and Grand Rivers, and Their Tributaries," Cleveland, May 22, 1968, Ohio WPCB Records, series 2397, box 19, pp. 338–44. Arnold W. Reitze, "Wastes, Water, and Wishful Thinking: The Battle of Lake Erie," *Case Western Reserve Law Review* 20, no. 1 (November 1968): 54 (quote).

Because these rivers do not cross state lines, under the 1965 act Ohio was not required to adopt water quality standards for these waters. However, under the Clean Water Restoration Act of 1966, states that adopted acceptable intrastate water quality standards were eligible for additional federal sewage treatment plant construction grant funds.

39. See, for example, Thomas L. Kimball, executive director of the National Wildlife Federation, letter to Udall, October 18, 1967, Michigan DNR Records, A/N

80-85, box 15, file 2, Michigan State Archives; and statement by Michigan United Conservation Clubs, in Transcript of Proceedings, MWRC Hearing, April 27, 1967, in Detroit, Romney Papers, box 337, "Water Quality Standards for St. Clair River, etc." file, pp. 58–59. Assistant secretary of Interior Max Edwards, letter to Representative John Dingell, March 5, 1968, Interior documents obtained under Freedom of Information Act, p. 2 (quote).

40. For example, Governor James Rhodes, letter to Udall, March 22, 1968 and Udall, letter to Rhodes, June 26, 1968, Rhodes Papers, manuscript series 353, box 8, file 18, and box 6, file 17, respectively. For an overview of the nondegradation controversy, see Holmes, *Programs and Policies,* pp. 189–90. For Interior approval, see Micklin, "Water Quality," p. 143.

41. Act of November 3, 1966, 80 *Statutes at Large,* p. 1249; and Holmes, *Programs and Policies,* p. 190.

42. Clarence W. Klassen, "Fitting a State Program to Federal Objectives," *Journal, WPCF* 40 (October 1968): 1702–10 (1702).

43. "Federal Pollution Program Knocked at ASCE Meeting," *Engineering News-Record,* October 26, 1967, pp. 18–19.

44. Joe Moore, "Water Quality Management in Transition," *Civil Engineering* 38 (June 1968): 30–32. In December 1967, James Quigley resigned as commissioner to take an executive position with a paper company.

45. *Lake Erie Conference, June 1966,* p. 527. In a public address at about the same time, G. A. Hall, secretary of the Ohio WPCB, defended the board's past accomplishments but admitted that state engineers had sometimes failed to provide a sufficient factor of safety when allocating stream assimilative capacity. See Hall, "Cleaner Water and Higher Goals," *Clean Waters for Ohio* 15, no. 2 (summer 1966): 12.

46. *Proceedings of the Conference in the Matter of Pollution of Lake Michigan and Its Tributary Basin, Chicago, January–March 1968* (Washington, D.C.: FWPCA, 1968) (hereafter *Lake Michigan Conference, January–March 1968*), p. 2984.

47. *Calumet Conference, February 1966,* p. 64.

48. Gargan, "Water Pollution in New York State," pp. 186–91, 208–13. Gargan notes that "in several interviews with members of the New York State bureaucracy, members of the Governor's staff, legislative staff and interest group leaders, a common response was that until 1965 no one was very much interested in water pollution as a problem" (pp. 210–11).

49. See testimony of Rockefeller in House Committee on Public Works, *Water Quality Act Hearings, February 1965,* pp. 224–28.

50. Ibid.

51. In his public statements in support of the Pure Waters program, Rockefeller downplayed industry's contribution to New York's water pollution problem, saying that New York industry had "done a pretty good job" in this area. *Lake Erie Conference, Buffalo, August 1965,* pp. 88–89 (88).

52. Gargan, "Water Pollution in New York State," pp. 260–64, 324–26, 434.

53. *Lake Erie Conference, Buffalo, August 1965,* pp.85–86.

54. "Reuther and Udall Agree on Need for Pure-Water Funds," *Detroit News,* November 7, 1965.

55. For disappointment within the administration over the legislation's provisions, see Joseph Califano, memo to President Johnson, September 16, 1966, WHCF, Legislative Background Files, Water Pollution, box 2, "1966 no. 3 of 3" file, Johnson Library;

and BOB director Charles Schultze, memo to President Johnson, October 15, 1966, WHCF, LE/HE 8-4, box 61, "LE/HE 8-4" file, Johnson Library. For budgets, see Holmes, *Programs and Policies,* p. 99.

56. Holmes, *Programs and Policies,* pp. 193–95.

57. "Voters OK Pollution, Park Bonds," *Detroit News,* November 6, 1968. The Michigan bond issue received widespread support from interest groups across the state. Many supporters emphasized the importance of the bond's passage to the state's billion-dollar tourist industry.

58. "Apathy Dooms State Bonds for Pollution," *Chicago Tribune,* November 7, 1968. Actually, the bond issue was approved by a substantial majority of those who voted on the proposition. The problem was that under the state constitution, the bond issue required a majority of the votes cast for members of the state assembly to pass. Apparently, many voters ignored this part of the ballot—an action equivalent to a no vote under these circumstances. The proposition received much more support in the Chicago metropolitan area than it did in downstate regions. In addition, more than half of the total bond issue was to be devoted to water resource management and recreation, so it is somewhat misleading to portray this as a defeat of an antipollution bond issue.

59. BOB director Charles Schultze had foreseen this problem when the Clean Water Restoration Act was first passed. In a memo to President Johnson, he predicted that the unrealistic spending levels authorized in the act would only generate disappointment and complaints about "cuts" when the administration submitted budgets with lower appropriations for the grant program. Schultze attached an article from *Nation's Cities* that had broken down the total yearly authorizations into the individual allocations that would be available to the states. Schultze, memo to Johnson, December 16, 1966, BOB Records, series 61.1, 1966, "Water Pollution" file.

In fairness to the states, it should be noted that prior to fiscal year 1968, the Kennedy and Johnson administrations never failed to request full funding of the construction grant program. See Holmes, *Programs and Policies,* p. 99.

60. "Clean Water Plan Fouled by Red Tape and Red Ink," *Detroit Free Press,* July 10, 1967.

61. *Proceedings, Progress Evaluation Meeting of the Conference in the Matter of Pollution of Lake Erie and Its Tributaries, Cleveland, June 4, 1968* (Washington, D.C.: FWPCA, 1968) (hereafter *Lake Erie Conference, June 1968*), pp. 76–77; and "Passage of Clean Water Issue Is Called a Challenge to City," *Cleveland Press,* November 6, 1968.

62. U.S. Congress, House Committee on Government Operations, *Water Pollution—Great Lakes (Part 3): Western Lake Erie, Detroit River, Lake St. Clair, and Tributaries: Hearings before the Subcommittee on Natural Resources and Power,* 89th Cong., 2d sess., September 9, 1966, pp. 1034–36.

63. "Report of the Task Force on Pollution Abatement," October 9, 1965, BOB Records, series 61.1, 1965, "Water Pollution" file, p. 24; and remarks by FWPCA Commissioner Quigley in "Industry Joins Battle to Stem Pollution Tide," *Business Week,* December 31, 1966.

While the federal task force recommended against special tax concessions for pollution abatement, critics of the federal program pointed out that federal contributions to municipal treatment plants that handled a large volume of industrial waste amounted to indirect taxpayer subsidies of industry. See Zwick and Benstock, *Water Wasteland,* pp. 322–31. Most of the Great Lakes states offered some special tax write-offs for industrial waste treatment facilities.

64. For the problems caused by municipal treatment of industrial wastes in New York, see New York Water Resources Commission Minutes, September 5, 1968, pp. 6–7, New York Department of Environmental Conservation, Albany. Industries were expected to pay equitable user fees and provide adequate pretreatment, if necessary, but this was often not the case.

65. For the Lake Ontario Basin, U.S. Congress, House Committee on Government Operations, *Water Pollution—Great Lakes (Part 1): Lake Ontario and Lake Erie: Hearings before the Subcommittee on Natural Resources and Power*, 89th Cong., 2d sess., July 22, 1966, pp. 128–30. For Lake Michigan, *Proceedings, Progress Evaluation Meeting in the Matter of Pollution of the Interstate Waters of the Grand Calumet River, Little Calumet River, Calumet River, Wolf Lake, Lake Michigan, and Their Tributaries, Chicago, September 11, 1967* (Washington, D.C.: FWPCA, 1967) (hereafter *Calumet Conference, September 1967*), pp. 180–81.

66. *Lake Erie Conference, June 1966*, pp. 29–43, 64.

67. For percentages, ibid., pp. 454, 473. For detergents see, for example, the comments by Gerald Remus in House Committee on Government Operations, *Water Pollution—Great Lakes (Part 3), September 1963*, p. 523.

68. Bueltman is cited from Lake Erie Enforcement Conference Technical Committee, Minutes, October 13–14, 1966, EPA Records, A/N 74-22, box 29, "Lake Erie, Admin." file no. 1. *Lake Erie Conference, March 1967*, p. 94 (quote).

69. U.S. Department of Interior press release, August 4, 1967, Ohio WPCB Records, series 1800, box 19, file 3. Membership in the task force was later expanded to include representatives from the DOA and the fertilizer industry.

70. Ibid.

71. *Proceedings, Technical Session of the Conference in the Matter of Pollution of Lake Erie and Its Tributaries, Cleveland, August 26, 1968* (Washington, D.C.: FWPCA, 1969) (hereafter *Lake Erie Conference, August 1968*), pp. 92–110.

72. *Lake Erie Conference, June 1968*, pp. 436–60.

73. George Harlow, memo to H. W. Poston, August 16, 196, EPA Records, A/N 74-22, box 29, "Lake Erie, Admin." file no. 2; and *Proceedings, Fourth Session of the Conference in the Matter of Pollution of Lake Erie and Its Tributaries, Cleveland, October 4, 1968* (Washington, D.C.: FWPCA, 1969) (hereafter *Lake Erie Conference, October 1968*), pp. 43–47, 63.

74. *Lake Erie Conference, August 1968*, pp. 59–61, 75–79.

75. *Lake Erie Conference, October 1968*, pp. 125–31.

76. *Calumet Conference, February 1966*, pp. 70–71.

77. George Romney, letter to Gaylord Nelson, April 22, 1966, and attached letters and memos, Romney Papers, box 336, "Water Pollution, Lake Michigan" file. Romney's advice to Knowles is in a letter dated April 14, 1966, in this file.

78. Robert Sobel and John Raimo, eds., *Biographical Directory of the Governors of the United States, 1789–1978* (Westport, Conn.: Meckler Books, 1978), pp. 1759–60; and Huffman, "Protectors of the Land and Water," pp. 537–40 (Huffman's study contains an extensive analysis of the Knowles-Nelson rivalry). Warren Knowles, letter to John Conway, September 2, 1965, Warren Knowles Papers, series 2142, box 18, file 39, Wisconsin State Archives; Knowles is quoted from his letter to H. R. Moore, president of Bergstrom Paper Company, July 18, 1966, box 15, file 42.

79. "U.S. Ready to Act on State Pollution," *Milwaukee Sentinel*, March 24, 1966; and "U.S. Pollution Official Criticized by Knowles," *Milwaukee Journal*, April 30, 1966.

80. Huffman, "Protectors of the Land and Water," pp. 482–84.

81. See, for an example of newspaper criticism, "Gov. Knowles Blocks Efforts to Clean Up Regional Waters," *Madison Capital Times*, April 2, 1966. For Lucey, see U.S. Congress, House Committee on Government Operations, *Water Pollution—Great Lakes (Part 4): Southwestern Lake Michigan: Milwaukee, Root, and Pike Rivers: Hearings before the Subcommittee on Natural Resources and Power*, 89th Cong., 2d sess., September 16, 1966, pp. 879–80. On Knowles's reaction and Udall's comment, see Huffman, "Protectors of the Land and Water," pp. 631–37. I discuss the Wisconsin reorganization along with those of other states in chapter 4.

82. Huffman, "Protectors of the Land and Water," p. 632.

83. Robert J. Schneider, "The Lake Michigan Enforcement Conference," in *Proceedings of the Twenty-Third Industrial Waste Conference in Lafayette, Indiana, May 7–9, 1968*, Purdue University Extension Series No. 132 (West Lafayette: Purdue University, 1968) (hereafter *Industrial Waste Conference, May 1968*), pp. 978–84. The alewife is not native to the Great Lakes, and the species' struggle to adjust to freshwater conditions seems to result in periodic die-offs. See William Ashworth, *The Late, Great Lakes* (New York: Alfred Knopf, 1986), pp. 120–21.

84. *Calumet Conference, September 1967*, pp. 181–90; and "Daley Defends Pollution Deadline," *Chicago Daily News*, September 13, 1967.

85. The telegram can be found in FWPCA Records, A/N NC3-382-81-2, box 23, "River Studies—Lake Michigan" file. "Pollution Demands U.S. Action," *Chicago Tribune*, September 29, 1967. The events leading up to the conference are summarized in Schneider, "The Lake Michigan Enforcement Conference," pp. 978–84.

86. *Lake Michigan Conference, January–March 1968*, pp. 98–111 (104).

87. Conference recommendations are summarized in Schneider, "The Lake Michigan Enforcement Conference," pp. 984–87.

88. *Lake Michigan Conference, January–March 1968*, pp. 590–97.

89. Ibid., pp. 2911–19. Several weeks before the opening of the conference, representatives from chambers of commerce, manufacturers associations, and other business groups from the states and cities bordering Lake Michigan met in Chicago and agreed to recommend formally that the federal government give the four state governments until December 31, 1972, to implement fully abatement programs contained in their interstate water quality standards before Washington took any independent action in the region. See ibid., pp. 1647–51.

90. Letters to Holmer from Green Bay Packaging, February 29, 1968; American Can, February 28, 1968; Charmin Paper, March 1, 1968; and Fort Howard Paper, March 1, 1968, Division of Environmental Protection Records, Wisconsin DNR, A/N 1989-257, box 2 (unprocessed)(Wisconsin State Archives) .

91. "Anti-Pollution Plans Challenged," *Wisconsin State Journal*, March 6, 1968; Schneider, "The Lake Michigan Enforcement Conference," pp. 984–85.

92. *Proceedings of the Second Session of the Conference in the Matter of Pollution of the Interstate Waters of the Grand Calumet River, Little Calumet River, Calumet River, Wolf Lake, Lake Michigan, and Their Tributaries, Chicago, December 11–12, 1968, and January 29, 1969* (Washington, D.C.: FWPCA, 1969) (hereafter *Calumet Conference, December 1968*), pp. 453–57. At the June 1968 session of the Lake Erie enforcement conference, George Harlow of the FWPCA Cleveland office reported that in most areas lake water quality had declined since 1963–1964. See *Lake Erie Conference, June 1968*, pp. 140–42, 147–48.

93. "Lake Erie Bathing Beach Survey," September 29, 1967; and "Statement of H. W. Poston, Chairman, U.S. Section, Lake Erie–Ontario Advisory Board," presented at IJC Hearing, January 16, 1968, Niagara Falls, New York, Executive Department Records, Interior, box 61, "FWPCA: Chicago Office" file, Johnson Library.

94. *Lake Erie Conference, June 1968,* pp. 17–20 (19). For similar sentiments, see the editorial, "Lake Cleanup Pressure Needed," *Cleveland Plain Dealer,* June 4, 1968. *Chicago Tribune,* September 12, 1967.

95. Rick Friedman, "Pictures Play Key Role in Winning a 'Pulitzer'," *Editor and Publisher,* May 13, 1967, p. 18; "A Special Magazine for 'Save Our Lake'," *Chicago Tribune,* November 24, 1967. Public officials interpreted the heavy newspaper coverage as a barometer of public concern. For example, the September 1967 meeting of the Conference of Great Lakes Senators was originally intended to focus on shipping issues, but the senators made pollution of the lakes the major topic. The leadoff speaker, Senator Charles Percy of Illinois, announced that "yesterday when I was in Chicago, pollution seemed to be on everyone's mind. . . . This morning, a leading Chicago newspaper filled three pages with an article discussing the problem of pollution. If pollution is the people's number one priority problem, it must be the number one priority of this conference." See *Chicago Tribune,* September 21, 1967.

96. *Cleveland Press,* May 2, 1966. For a very critical discussion of the federal officials' willingness to compromise with dischargers, see Zwick and Benstock, *Water Wasteland,* chaps. 10, 13.

97. Peter Kuh, memo to Stein, March 31, 1966, EPA Records, A/N 74-22, box 47, "Lake Michigan (Calumet), Admin." file no. 1.

98. *Proceedings, Progress Meeting in the Matter of Pollution of the Interstate Waters of the Grand Calumet River, Little Calumet River, Calumet River, Wolf Lake, Lake Michigan, and Their Tributaries, Chicago, March 15, 1967* (Washington, D.C.: FWPCA, 1967) (hereafter *Calumet Conference, March 1967*), pp. 336, 452–55, 840–43.

99. "Daley Defends Pollution Deadline," *Chicago Daily News,* September 13, 1967; "Anti-Wastes Date Kept for Steel Firms," *Chicago Tribune,* September 20, 1967.

100. "Three Steel Firms Fail to Convince Lake Probers," *Chicago Sun-Times,* October 27, 1967.

101. Bacon quoted from "U. S., Indiana Split on Deadline for Ending Lake Pollution," *Chicago Sun-Times,* September 29, 1967, and from "Lawyers Leap into Dirty Battle," *Engineering News-Record,* October 26, 1967, p. 17.

102. Allen Lavin, letter to Illinois attorney general William G. Clark, October 24, 1967, EPA Records, A/N 74-22, box 48, "Lake Michigan (Calumet), Litig." file no. 1; and "Sanitary District Sues U.S. Steel over Calumet River Pollution," *Chicago Daily News,* October 26, 1967.

103. Lavin, "Energetic Enforcement of Industrial Waste Ordinances," in *Industrial Waste Conference, May 1968,* pp. 550–53.

104. Ibid. The Chicago MSD suit was not fully resolved until the late 1970s. "Sanitary District Sues U.S. Steel over Calumet River Pollution," *Chicago Daily News,* October 26, 1967; *Lake Michigan Conference, January–March 1968,* pp. 360–61.

105. Save the Dunes representative cited from *Lake Michigan Conference, January–March 1968,* pp. 2037–40. For similar sentiments, see the testimony of the League of Women Voters of Indiana, pp. 1426–28; the Illinois Federation of Sportsmen's Clubs, pp. 1722–25; and the Wisconsin Resources Conservation Council, pp. 2877–78. Stein cited from *Calumet Conference, December 1968,* pp. 223–24.

106. *Calumet Conference, December 1968*, pp. 500–510. Along with the three Indiana steel firms and U.S. Steel South Works, a number of plants within the MSD discharging to the Calumet River also requested extensions. Wastes discharging into the Calumet River almost always ended up in the Illinois River system that received the effluent from the MSD. There was considerable dispute at the conference over whether these plants were covered by the conference.

107. Ibid., pp. 725–34.

108. Gladwin Hill, "Water Seeks a Higher Level," *Water and Wastes Engineering* 4 (February 1967): 60. Katz cited from Transcript of Proceedings, Indiana SPCB Hearing, April 10, 1967, in Gary, Indiana, Indiana SPCB Records, River Basin files, box R2258, "Lake Michigan Basin" file, pp. 9–10. See Hurley, "Environmental and Social Change in Gary, Indiana," for an extended account of environmental politics in Gary during this era. In general, see many of the comments by civic groups and conservation organizations in the proceedings of the federal enforcement conferences and the transcripts of the state hearings on water quality standards.

109. Reitze, "The Battle of Lake Erie," pp. 61–62, 66.

Chapter 4: The New Regulation

1. Thad L. Beyle, Thomas E. Peddicord, and Francis H. Parker, *Integration and Coordination of State Environmental Programs* (Lexington, Ky.: Council of State Governments, 1975), pp. 26–28.

2. Ibid., pp. 15–16. In Elizabeth H. Haskell and Victoria S. Price, *State Environmental Management: Case Studies of Nine States* (New York: Praeger, 1973), ch. 1 examines Illinois, ch. 2 examines Minnesota, ch. 4 Wisconsin, ch. 5 New York, and ch. 9 Michigan.

3. Larry Sabato, *Goodbye to Good-Time Charlie: The American Governor Transformed, 1950–1975* (Lexington, Mass.: Lexington Books, D. C. Heath and Company, 1978).

4. Riley E. Dunlap, "Trends in Public Opinion toward Environmental Issues, 1965–1990," in *American Environmentalism: The U.S. Environmental Movement, 1970–1990*, ed. Dunlap and Angela G. Mertig (New York: Taylor and Francis, 1992), pp. 92–93.

5. Nelson A. Rockefeller, *Our Environment Can Be Saved* (New York: Doubleday, 1970); and "Excerpts from Remarks by Governor Milliken at Environmental Teach-Ins," April 22, 1970, Milliken Papers, box 1224, "Environment, 1971" file.

6. Beyle et al., *State Environmental Programs*, pp. 26–28. In Indiana, the SPCB continued to regulate water pollution. In Michigan, the MWRC remained intact, although it was absorbed into the Michigan DNR and the Natural Resources Commission assumed some of the MWRC's decision-making authority.

7. Ibid.

8. Wilson, "Politics of Regulation," pp. 386–87. For a sample of such criticism from the Great Lakes region, see the testimony of a representative from the Detroit-area Downriver Antipollution League in *Proceedings, Fifth Session of the Conference in the Matter of Pollution of Lake Erie and Its Tributaries, Detroit, June 3–4, 1970* (Washington, D.C.: FWQA, 1971) (hereafter *Lake Erie Conference, June 1970*), pp. 508–12.

9. Purdy, "Communication Techniques in Water Pollution Control: State Pro-

grams," *Journal, WPCF* 42 (January 1970): 15; and *New York Times,* December 7, 1970. Purdy noted with pleasure that each successive member on the MWRC representing conservation groups quickly grasped the complexities and difficulties of pollution abatement and acted as a moderating influence on the views of clean water advocates.

10. See Haskell and Price, *State Environmental Management.* For the tradition of citizen boards, see also McConnel, *Private Power;* for the role of citizen policy boards in the new environmental agencies, see also Beyle et al., *State Environmental Programs,* pp. 42–43.

11. See *Report of the Citizens Task Force on Environmental Protection* (State of Ohio, 1971), p. 153.

12. Haskell and Price, *State Environmental Management,* pp. 51, 60–61; and Thomas F. Bastow, *"This Vast Pollution . . .": United States of America v. Reserve Mining Company* (Washington, D.C.: Green Fields Books, 1986), pp. 57–58.

13. Executive Office press release, June 29, 1970, Richard Ogilvie Papers, series 101.37, "Environmental Quality Reorganization" file, Illinois State Archives.

14. Haskell and Price, *State Environmental Management,* ch. 1.

15. "Environmental Watchdog for State," *New York Times,* April 24, 1970; Diamond biography in *New York Red Book, 1970–1971,* p. 570; and Haskell and Price, *State Environmental Management,* pp. 144, 158–59.

16. Haskell and Price, *State Environmental Management,* pp. 27–28.

17. For Nixon's personal view of environmental issues, see Ruckelshaus *Oral History Interview,* pp. 10–11, 36–37; and EPA *Oral History Interview* No. 2, Russell E. Train, July 1993 (hereafter Train *Oral History Interview*), pp. 7, 13–14. For an overview of Nixon's environmental policy by an administration insider, see John C. Whitaker, *Striking a Balance: Environment and Natural Resources Policy in the Nixon-Ford Years* (Washington, D.C.: American Enterprise Institute for Public Policy Research, 1976).

18. Quarles, *Cleaning Up America,* ch. 1.

19. "New Team for Water Cleanup," *Chemical Week,* May 31, 1969, pp. 61–62; Zwick and Benstock, *Water Wasteland,* pp. 60–65; and Quarles, *Cleaning Up America,* pp. 143–44. Dominick was a cousin of Republican senator Peter Dominick of Colorado and had served as legislative assistant to Republican senator Clifford Hansen of Wyoming.

20. "New Team for Water Cleanup," *Chemical Week,* May 31, 1969; "Carl Klein Takes Hold with a Cast Iron Grip," *Engineering News-Record,* August 28, 1969, p. 41; and Zwick and Benstock, *Water Wasteland,* p. 61.

21. Alfred A. Marcus, *Promise and Performance: Choosing and Implementing an Environmental Policy* (Westport, Conn.: Greenwood Press, 1980), ch. 1. See also Whitaker, *Striking a Balance,* ch. 3.

22. "William D. Ruckelshaus Biography," in Ruckelshaus *Oral History Interview,* p. iv.

23. Ibid., p. v.

24. Ibid., p. 9; and Quarles, *Cleaning Up America,* pp. 27–28, 36. I am indebted to Dennis Williams, former historian at EPA, for insight into Ruckelshaus's legacy at the agency.

25. *Proceedings, Reconvened Second Session of the Conference in the Matter of Pollution of the Interstate Waters of the Grand Calumet River, Little Calumet River, Calumet River, Wolf Lake, Lake Michigan, and Their Tributaries, Chicago, August 26, 1969* (Washington, D.C.: FWPCA, 1970) (hereafter *Calumet Conference, August 1969*), pp. 59–61, 67, 74, 264.

26. *Proceedings, Progress Evaluation Meeting of the Conference in the Matter of Pollution of Lake Erie and Its Tributaries, Cleveland, June 27, 1969* (Washington, D.C.: FW-PCA, 1970) (hereafter *Lake Erie Conference, June 1969*), pp. 446–53, 534.

27. Ibid., pp. 9–17, 537–38.

28. O'Brien, "Environmentalism as a Mass Movement," pp. 17–19. See also Sale, *The Green Revolution,* pp. 18–24; Mitchell, "From Conservation to Environmental Movement," pp. 81–113; and Gottlieb, *Forcing the Spring,* pp. 133–43.

29. Hurley, "Environmental and Social Change in Gary, Indiana," pp. 113–14.

30. Robert V. Bartlett, *The Reserve Mining Controversy: Science, Technology, and Environmental Quality* (Bloomington: Indiana University Press, 1980), p. 68; and Bastow, *"This Vast Pollution . . . , "* p. 31.

31. Hurley, "Environmental and Social Change in Gary, Indiana," pp. 114–16. The quote is from the testimony of Patricia Kaltwasser of Housewives to End Pollution, in *Lake Erie Conference, June 1970,* p. 418. These women were acting in a tradition of environmental activism established by an earlier generation of women's groups. For the role of women and women's groups in the conservation and sanitary reform movements, see Carolyn Merchant, "Women of the Progressive Conservation Movement: 1900–1916," *Environmental Review* 8, no. 1 (spring 1984): 57–85; Suellen M. Hoy, "'Municipal Housekeeping': The Role of Women in Improving Urban Sanitation Practices, 1880–1917," in Melosi, *Pollution and Reform,* 173–98; and Martin V. Melosi, *Garbage in the Cities: Refuse, Reform, and the Environment, 1880–1980* (College Station: Texas A&M University Press, 1981), pp. 117–24.

32. *Lake Erie Conference, June 1970,* pp. 405–9.

33. See, for example, the comments of Joseph Karaganis of BPI, in *Calumet Conference, August 1969,* pp. 30–32. Karaganis compared corporate polluters to narcotics dealers in terms of their destructive impact on society.

34. Heclo, "The Sixties' False Dawn," pp. 34–63.

35. Leonard Silk and David Vogel, *Ethics and Profits: The Crisis of Confidence in American Business* (New York: Simon and Schuster, 1976), p. 21. For a good discussion of changing public attitudes toward corporations and the public interest movement's efforts to force changes in corporate behavior, see David Vogel, *Lobbying the Corporation: Citizen Challenges to Business Authority* (New York: Basic Books, 1978), especially pp. 3–29.

36. Vogel, "The Public-Interest Movement," pp. 607–27.

37. An excellent account of the founding and early activities of the Environmental Defense Fund can be found in Thomas R. Dunlap, *DDT: Scientists, Citizens, and Public Policy* (Princeton: Princeton University Press, 1981), especially pp. 143–54. See also Mitchell, "From Conservation to Environmental Movement," pp. 88–92; and Gottlieb, *Forcing the Spring,* pp. 136–43.

38. "Business and the Public Interest," *New York Times,* December 19, 1971.

39. Senate Committee on Public Works, *Water Pollution (Part 2), June 1965,* pp. 903–4; and "Action Urged on Pollution," *Toledo Blade,* May 27, 1965.

40. Mary Ann Heidemann, "Regional Ecology and Regulatory Federalism: Wisconsin's Quandary over Toxic Contamination of Green Bay" (Ph.D. diss., University of Wisconsin–Madison, 1989), pp. 335–36; and "Top U.S. Antipollution Agent Urges Intense Citizen Action," *Chicago Sun-Times,* January 13, 1971.

41. The text of this act can be found in Arnold W. Reitze Jr., *Environmental Law,* 2d ed. (Washington, D.C.: North American International, 1972), vol. 1, pp. 52–54.

42. Joseph L. Sax, *Defending the Environment* (New York: Alfred A. Knopf, 1971).

The Sax Act did not lead to a flood of litigation, as some feared. Only thirty-six suits were filed under the law in its first seventeen months on the books. Corporations were defendants in thirteen of the cases, while state and local agencies were the targets in about half the cases. Surprisingly, government agencies used the new act to bring suit or intervene in one-third of the cases. Most of the cases were settled out of court. See Haskell and Price, *State Environmental Management,* ch. 9.

43. Raymond C. Hubley Jr., Izaac Walton League executive director, letter addressed to "Dear League Leader," February 25, 1972, Izaac Walton League Records, manuscript series 882, box 3, folder 4, Ohio Historical Society, Columbus. A copy of the resolution creating the legal defense fund is attached to the letter.

44. "Citizen Help in Fighting Polluters Hailed by Scott," *Chicago Tribune,* July 20, 1972; and *Chicago Tribune,* August 13, 1972.

45. "38 Percent of Lake Ontario Polluters in State Are Lagging in Cleanup," *Buffalo Evening News,* March 3, 1971.

46. *Proceedings, Third Session of the Conference in the Matter of Pollution of Lake Michigan and Its Tributary Basin, Milwaukee, March 31 and April 1, 1970* (Washington, D.C.: FWQA, 1970) (hereafter *Lake Michigan Conference, March–April 1970*), pp. 296–97. For similar views from Klassen's New York counterpart, see "Metzler Leads a Pure Water War," *Engineering News-Record,* September 11, 1969, pp. 31–32.

47. New York State Department of Environmental Conservation, Division of Pure Waters, "Report to International Joint Commission Advisory Board—Lake Erie," August 31, 1971, EnCon Records, series A1118, box 7, "General Correspondence, 1968–1972" file. In 1963, the Illinois Sanitary Water Board referred two cases to the Illinois attorney general's office; in 1964, it referred four cases; in 1965, fourteen; in 1966, eleven; in 1967, fifteen; in 1968, twenty-three; in 1969, fifty-seven; and in 1970, seventy-one. See Michael Orin Ayers, "An Evaluation of Water Pollution Control Arrangements in and by the State of Illinois" (Ph.D. diss., University of Oklahoma, 1974), p. 99.

48. "Statement of Attorney General Robert Warren to Wisconsin Natural Resources Board," February 12, 1970, Knowles Papers, series 2142, box 31, file 33.

49. *Illinois Attorney General's Report, 1969–1970* (Springfield: State of Illinois, 1971) pp. xx–xxi; and *Calumet Conference, August 1969,* pp. 25–27. For biographical information on Scott, see *Illinois Blue Book, 1979–1980,* p. 22.

50. *Calumet Conference, August 1969,* pp. 14–17; and "A Sorry Performance," *Chicago Tribune,* August 28, 1969 (quote). For similar disappointment, see "Clean Up the Air and Water," an editorial in the *Chicago Sun-Times,* August 31, 1969.

51. James S. Cannon and Jean M. Halloran, *Environmental Steel: Pollution in the Iron and Steel Industry* (New York: Praeger, 1974), pp. 459–60; and Clarence Klassen, letter to Secretary Hickel, November 18, 1969, EPA Records, A/N 75-63, box 2, "Lake Michigan (Calumet) Legal" file. U.S. Steel South Works discharged directly to Lake Michigan. Except in unusual circumstances, the effluent from these other sources was diverted to the Illinois River–Mississippi River system.

52. "Illinois Governor Urges Suit against Three States," *Wall Street Journal,* October 27, 1969 (quote); "Remarks of Governor Ogilvie, Civic Federation Luncheon," October 24, 1969, pp. 8–12, and UPI wire story, October 24, 1969, Milliken Papers, box 757, "WRC, General" file.

53. "Pollution Control: Murkier and Murkier," *Iron Age,* July 9, 1970, p. 50.

54. *Lake Michigan Conference, March–April 1970,* pp. 310–11; *Proceedings, Executive Session of the Third Session of the Conference in the Matter of Pollution of Lake Michigan*

and Its Tributary Basin, Chicago, May 7, 1970 (Washington, D.C.: FWQA, 1970) (hereafter *Lake Michigan Conference, May 1970*), pp. 46–57, 84–88; and *Illinois Attorney General's Report, 1969–1970*, pp. xxii.

55. "U.S. Steel Yields on Lake Dumping," *New York Times,* January 24, 1971.

56. Cannon and Halloran, *Environmental Steel,* p. 460.

57. Ibid., p. 280; and *Illinois Attorney General's Report, 1973–1974,* p. 11.

58. "Five Billion Expected to Clean Water," *Chicago Tribune,* February 20, 1970; and *Illinois Attorney General's Report, 1969–1970,* p. xxii.

59. Glenn D. Pratt, memo to State Program Officer, June 12, 1972, EPA Records, A/N 73-43, box 2, "Region 5" file. For firsthand accounts of his office's activities in this area, see the texts of Brown's speeches on environmental issues, Ohio Attorney General William J. Brown Records, series 1816, box 1, Ohio State Archives.

60. *New York Times,* October 30, 1970; and "Lefkowitz Bids Court Close Polluting Refinery," *New York Times,* October 27, 1970.

61. The individual actions are summarized by the Interior Department in U.S. Congress, Senate Committee on Public Works, *Water Pollution 1970 (Part 1): Hearings before the Subcommittee on Air and Water Pollution,* 91st Cong., 2d sess., April 20–21, 27, 1970, pp. 333–40. Hickel also sent a notice of violation to an Eagle-Pitcher Industries plant in Kansas the same day.

62. Ibid.; and Harlow, memo to H. W. Poston, October 10, 1969, EPA Records, A/N 74-22, box 29, "Lake Erie, Admin." file no. 2.

63. "Statement of Dr. Emmet Arnold, FWPCA Conference with J&L Steel," October 8, 1969, Ohio WPCB Records, series 1800, box 1, file 26, p. 1; and "Rhodes Raps U.S. Methods on Pollution," *Cleveland Plain Dealer,* October 1, 1969.

64. Whitaker, *Striking a Balance,* pp. 36–37; and Holmes, *Programs and Policies,* pp. 103–4.

65. Albert E. Cowdrey, "Pioneering Environmental Law: The Army Corps of Engineers and the Refuse Act," *Pacific Historical Review* 44 (August 1975): 345.

66. "Pollution Laid to Eleven Companies," *New York Times,* February 10, 1970; and Allan G. Kirk II, associate solicitor, Water Resources and Procurement, letter to Walter Kiechel Jr., Department of Justice, September 24, 1969, EPA Records, A/N 74-22, box 47, "Lake Michigan (Calumet), Admin." file no. 2. Observers also noted that Foran was a Democratic holdover who had not yet been replaced at Justice, whereas Scott was a Republican.

67. "Pollution Laid to Eleven Companies," *New York Times,* February 10, 1970. The waste from these firms entered the Illinois River–Mississippi River system, not the Lake Michigan Basin. For changes in the Corps, see Cowdrey, "Pioneering Environmental Law," p. 346. For late February suits, see Cannon and Halloran, *Environmental Steel,* p. 433. For federal grand jury, see "Cite U.S. Steel for Polluting Lake Michigan," *Chicago Tribune,* March 19, 1970, and Cannon and Halloran, *Environmental Steel,* p. 460.

68. Quarles, *Cleaning Up America,* pp. 7–8.

69. The mercury problem is discussed in more detail in chapter 5.

70. For brief profiles of Reuss, see "The Patient Patrician," *Time,* February 21, 1972, p. 24; "Reuss: Harbinger in the House," *Business Week,* September 18, 1971, p. 18; and Quarles, *Cleaning Up America,* pp. 69–70.

71. Cowdrey, "Pioneering Environmental Law," p. 346; Odom Fanning, *Man and His Environment: Citizen Action* (New York: Harper and Row, 1975), p. 77; and "U.S. Judge Dismisses Water Pollution Suits Brought by Rep. Reuss," *Wall Street Journal,* Feb-

ruary 25, 1971. Reuss turned his award over to the Wisconsin DNR to help fund sewage treatment plant construction.

72. Marcus, *Promise and Performance*, pp. 88–89, 98; "U.S. Will Drop Air and Water Pollution Control Conferences," *Chicago Tribune*, February 15, 1971; and "Uncle Sam's Lawyers: A Growing Force on the Pollution Scene," *Environmental Science and Technology* 5 (October 1971): 994–95.

73. Ruckelshaus, memo to John Whitaker, January 11, 1971, Staff Member and Office Files: John Whitaker, Nixon Presidential Materials Staff (hereafter Whitaker Files), box 135, "Great Lakes Agreement" file no. 2, National Archives; Ruckelshaus, memo to Whitaker, not dated but circa summer 1971, Whitaker Files, box 70, "Great Lakes " file no. 1; Representative Jack Kemp, letter to Shultz, November 4, 1971, and Don Crabill, memo to Bill Gifford, March 22, 1972, BOB Records, series 69.1, Subject Files of Director, 1970–1972, box 97, "Water Pollution" file. The specific provisions of the plan shifted over time, reflecting criticism from other administration officials.

74. *Lake Erie Conference, June 1970*, pp. 134–35, 142–43.

75. Ibid.; and *Proceedings, Third Session (Reconvened) of the Conference in the Matter of Pollution of Lake Michigan and Its Tributary Basin, Chicago, March 23–25, 1971* (Washington, D.C.: EPA, 1971) (hereafter *Lake Michigan Conference, March 1971*), pp. 889–95.

76. See, for example, "U.S. Warns 78 Cities, 44 Firms on Delaying Lake Erie Cleanup," *Cleveland Plain Dealer*, June 4, 1970; and Stein, memo to David Dominick, October 9, 1970, EPA Records, A/N 74-22, box 30, "Lake Erie, Admin." file no. 3. The FWPCA changed its name to the Federal Water Quality Administration (FWQA) in 1970 before being absorbed into the EPA. David Dominick became acting commissioner of the Water Quality Office, but Carl Klein resigned his position at Interior in September 1970.

77. Stein, memo to Dominick, February 3, 1971, EPA Records, A/N 74-22, box 30, "Lake Erie, Admin." file no. 3. See also "Uncle Sam's Lawyers."

78. See the summary statistics in EPA, *The First Two Years: A Review of EPA's Enforcement Program, February 1973* (Washington, D.C.: U.S. EPA, 1973), pp. 11–17. These are primarily aggregate statistics for the entire nation. As of September 1972, the Region 5 Chicago Office, which possessed primary responsibility for the Great Lakes, had been involved in fifty-nine 180-day notices, seventy-eight Refuse Act civil cases (sixty-two initiated by the Justice Department), and fifty-nine Refuse Act criminal cases. This total included actions in other waters within the six-state region of Minnesota, Wisconsin, Illinois, Indiana, Michigan, and Ohio, but a large number of these cases involved dischargers in the Great Lakes Basin. See George Marienthal, director, EPA Office of Regional Liaison, "Report on Visit to Chicago Regional Office," September 12, 1972, EPA Records, A/N 74-30, box 3, "Region 5" folder no. 2, p. 3. See the listing of individual enforcement cases in EPA, *The First Two Years*, ch. 4.

79. "Old Law, New Charges," *Chemical Week*, March 3, 1971, p. 16; and "Sierra Club Prods Clean-Water Fight," *Cleveland Plain Dealer*, December 19, 1970.

80. Indiana SPCB Minutes, Book 41, October 17, 1972, p. 171; "U.S. Files Suit to Halt J&L Pollution of River," *Cleveland Plain Dealer*, December 19, 1970; and Cannon and Halloran, *Environmental Steel*, p. 298.

81. Criminal cases under the Refuse Act also won some penalties from the offending companies, but these relatively small fines were little more than an annoyance for most firms. For example, shortly after the Justice Department filed civil suit against Jones and Laughlin, the department also charged the firm with ten separate discharge violations

over a ten-day period in September 1970. Five of the charges were dismissed in court. The steel company pleaded no contest on the remaining five counts and was fined $5,000. See Cannon and Halloran, *Environmental Steel,* p. 298.

82. U.S. Congress, House Committee on Public Works, *Water Pollution Control Legislation—1971 (Oversight of Existing Program),* 92d Cong., 1st sess., May–July 1971, p. 241. See also "Too Many Cooks: Companies Complain That Pollution Laws Conflict, Change Often," *Wall Street Journal,* December 23, 1970.

83. Speer is quoted in "Emotion, Politics Fog Pollution Problems," *Chemical and Engineering News,* November 3, 1969, p. 25. "Is There Still Time to Save U.S. Industry?" *Industry Week,* October 4, 1971, pp. S9–S10; and Joseph F. Lagnese Jr., "Water Pollution Control Policy: A Need for Engineer Involvement," *Professional Engineer* (March 1972): 15–18.

84. Cannon and Halloran, *Environmental Steel,* pp. 16–17; and Gregg Kerlin and Daniel Rabovsky, *Cracking Down: Oil Refining and Pollution Control* (New York: Council on Economic Priorities, 1975), ch. 2.

85. Marcus, *Promise and Performance,* pp. 144–46; EPA, *The First Two Years,* pp. 4–5; and Timothy Atkeson, general counsel, memo to Russell Train, CEQ chairman, July 29, 1971, Whitaker Files, box 144, "Permit Program" file. In many of the Refuse Act civil cases, the negotiations and pretrial work had hardly begun when Congress enacted the Water Pollution Control Act Amendments of 1972 in October of that year. After that, many of the cases were held in abeyance until pollution control requirements could be established under the new procedures established by the amendments. See Quarles, *Cleaning Up America,* pp. 50–51.

86. Timothy Atkeson, general counsel, memo to Russell Train, CEQ chairman, July 29, 1971, Whitaker Files, box 144, "Permit Program" file.

87. Text of executive order, White House press release, December 23, 1970, Whitaker Files, box 98, "Refuse Act Permit" file no. 2; and *Environmental Quality: The Second Annual Report of the Council on Environmental Quality* (Washington, D.C.: GPO, 1971), pp. 10–11. Federal officials viewed the collection of detailed company effluent discharge data as another advantage of the program. No one wanted a replay of the frantic scramble to find industrial users of mercury that followed the discovery of high mercury concentrations in Lake St. Clair fish in the spring of 1970. See memo, Timothy Atkeson, general counsel, to Russell Train, CEQ chairman, July 29, 1971, Whitaker Files, box 144, "Permit Program" file.

88. Zwick and Benstock, *Water Wasteland,* pp. 290–96; Quarles, *Cleaning Up America,* pp. 103–4; and U.S. Congress, House Committee on Commerce, *Refuse Act Permit Program: Hearings before the Subcommittee on the Environment,* 92d Cong., 1st sess., February 18, 19, 1971 (hereafter *Refuse Act Permit Program Hearings*).

89. *Refuse Act Permit Program Hearings,* pp. 59–62, 92–95 (quote is from p. 59). The March 24, 1971, letter to Ruckelshaus is printed in *Lake Michigan Conference, March 1971,* pp. 485–87.

90. "All Waste Discharges in Nonnavigable Waters Ruled Illegal by Court," *Wall Street Journal,* December 27, 1971; Quarles, *Cleaning Up America,* pp. 110–11.

91. "All Waste Discharges in Nonnavigable Waters Ruled Illegal by Court," *Wall Street Journal,* December 27, 1971; Quarles, *Cleaning Up America,* pp. 110–11.

92. Ruckelshaus quoted from Ruckelshaus *Oral History Interview,* pp. 19–20; complaints from the testimony of Wesley Gilbertson, deputy secretary of the Pennsylvania Department of Environmental Resources, in House Committee on Public Works, *Water Pollution Control Legislation—1971,* p. 397. For similar sentiments, see the testi-

mony of Ralph Purdy, executive secretary of the MWRC, ibid., pp. 468–69. Eagle, letter to Stein, June 10, 1970, EPA Records, A/N 75-63, box 2, "Lake Erie, Legal" file no. 2.

93. Perry Miller, letter to Senator Birch Bayh Jr., September 21, 1971, Edgar Whitcomb Papers, Departmental Correspondence Files, 1969–1972, box 5, "Pollution" file, Indiana State Archives. Miller succeeded Blucher Poole as executive secretary in 1970. For the general state opposition to the Refuse Act permit program, see the various testimonies in House Committee on Public Works, *Water Pollution Control Legislation—1971.*

94. I discuss the major features of the law in detail in the epilogue.

95. "Collapse of 'Pure Waters' Laid Solely to Congress," *Buffalo Evening News,* January 22, 1972.

96. Richard M. Bernard, ed., *Snowbelt Cities: Metropolitan Politics in the Northeast and Midwest since World War II* (Bloomington: Indiana University Press, 1990).

97. Quarles, *Cleaning Up America,* pp. 37–38; Purdy's testimony to House Committee on Public Works, *Water Pollution Control Legislation—1971,* pp. 497–500.

98. See testimony of Stokes in U.S. Congress, Senate Committee on Public Works, *Water Pollution 1970 (Part 2): Hearings before the Subcommittee on Air and Water Pollution,* 91st Cong., 2d sess., April–May 1970, pp. 407–9.

99. Remus's figures are from *Lake Erie Conference, June 1969,* pp. 83–84. Detroit also benefited from a much more comprehensive state grant program that prefinanced the federal contribution.

100. "Pollution Pacts Reached by EPA, Three Major Cities," *Wall Street Journal,* June 1, 1971.

101. The Cleveland situation was complex, but negotiations are summarized succinctly in EPA, *The First Two Years,* pp. 30–33.

Chapter 5: The Burden of Proof

1. Mark Hollis, "Water Resources and Needs for Pollution Control," *Journal, WPCF* 32 (March 1960): 228–29 (229). Joel Tarr has noted the impact on pollution control policy of "the emerging discipline of environmental health, with its focus on the chronic, degenerative diseases. . . . In this regard, we have returned to the original thrust of the 19th century sanitary movement towards cleaning the urban environment in order to insure freedom from epidemics." See Tarr, "Historical Perspectives on Hazardous Wastes in the United States," *Waste Management and Research* 3 (1985): 101.

2. *Lake Michigan Conference, January–March 1968,* pp. 2264–68 (MacMullen), 2297–303 (Ball).

3. Lake Michigan Enforcement Conference Pesticides Committee, "Report on Insecticides in Lake Michigan," November 1968, pp. 1–10, Milliken Papers, box 756, "Pesticides, General" file. The committee focused on insecticides, since there was no evidence to suggest significant levels of pesticides other than insecticides in the lake.

4. Ibid., pp. 22–24.

5. Michigan Department of Agriculture, "DDT in Coho Salmon: Meeting with Dept. of Health, Education and Welfare of April 1, 1969," Milliken Papers, box 756, "Pesticides, General" file; and Dale Ball, memo to Milliken, April 11, 1969, and attached summary of meeting with HEW officials on April 10, 1969, Milliken Papers, box 756, "5 State Pesticide Conference" file.

6. "Great Lakes States Governors' Conference on Pesticides," April 20, 1969, and

"Recommendations of the Five State Governors' Conference on Pesticides," August 27, 1969, Milliken Papers, box 756, "Pesticides, Recommendations" file; and Dunlap, *DDT,* pp. 178–79.

7. Charles C. Edwards, letter to Kenneth Wilcox Jr., Michigan Department of Public Health, June 16, 1970; Dale Ball, memo to James Kellog, May 6, 1971; and Mark Mason, memo to Kellog, September 3, 1971, all in Milliken Papers, box 756, "DDT" file. Mason and Kellog were members of Governor Milliken's staff.

Michigan's coho salmon stocking program was an attempt to replace the lake trout, which had almost been wiped out by the sea lamprey, a species not native to the Great Lakes. State authorities hoped that anglers attracted to Michigan by the salmon would pump millions of dollars each year into the state tourism industry. See Dunlap, *DDT,* pp. 178–79.

8. Huffman, "Protectors of the Land and Water," pp. 808–9. For a detailed narrative of the movement to ban DDT use, with much information on events in Michigan and Wisconsin, see Dunlap, *DDT.*

9. William G. Turney, "Mercury Pollution: Michigan's Action Program," *Journal, WPCF* 43 (July 1971): 1429. For an overview of the mercury problem in the Great Lakes, see this entire article and the chronology of events in the "FWPCA Report on Mercury in the Lake St. Clair–Lake Erie Systems," in *Lake Erie Conference, June 1970,* following p. 724.

10. For background on the Minamata tragedy, see Patricia A. D'Itri and Frank M. D'Itri, *Mercury Contamination: A Human Tragedy* (New York: John Wiley and Sons, 1977), ch. 4. For industrial ignorance about the environmental dangers of mercury, see the testimony of Dow vice president Ben Branch in U.S. Congress, Senate Committee on Commerce, *Effects of Mercury on Man and the Environment (Part 1): Hearings before the Subcommittee on Energy, Natural Resources, and the Environment,* 91st Cong., 2d. sess., May 8, 1970, pp. 43–44, 54–55. D'Itri and D'Itri make a good case that both Canadian and American authorities should have been alerted much sooner to the dangers of mercury in the environment. Widely read scientific journals published articles on the Minamata tragedy as early as the late 1950s; Swedish scientists published reports on the natural conversion in waterways of inorganic mercury from industrial uses into methylmecury. See *Mercury Contamination,* pp. 56–57.

11. "Mercury Spills Imperil Erie Fisheries," *New York Times,* May 11, 1970; and D'Itri and D'Itri, *Mercury Contamination,* pp. 60–61.

12. "Chlorine Makers Clutch at Last Drops of Mercury," *Chemical Week,* February 24, 1971, p. 75.

13. Dr. Kenneth Wilcox Jr., letter to James Kellog, January 15, 1971, Milliken Papers, box 756, "DDT" file; "EPA Weekly Activities Report," October 4, 1971, EPA Records, A/N 74-30, box 1, "Weekly Reports" file; and *Third Annual Report of the Council on Environmental Quality* (Washington, D.C.: U.S. GPO, 1972), p. 127.

14. My summary of events in this section is drawn from Bartlett, *The Reserve Mining Controversy,* and Bastow, *"This Vast Pollution. . . ."* The former is a comprehensive and balanced study by a political scientist, the latter is a lively account by a federal lawyer who took part in the case.

15. William Simon Rukeyser, "Fact and Foam in the Row over Phosphates," *Fortune* (January 1972): 72–73; and William McGucken, *Biodegradable: Detergents and the Environment* (College Station: Texas A&M University Press, 1991), pp. 16–18.

16. U.S. Congress, House Committee on Government Operations, *Phosphates in*

Detergents and the Eutrophication of America's Waters: Hearings before the Subcommittee on Conservation and Natural Resources, 91st Cong., 1st sess., December 15–16, 1969 (hereafter *Phosphate Hearings, December 1969*), pp. 66–67. For a listing of the member companies of the Soap and Detergent Association, see p. 75.

17. For a detailed account of the detergent foaming problem and its resolution, see McGucken, *Biodegradable.*

18. McGucken concludes that the industry would have developed a biodegradable surfactant regardless of political pressures, although he concedes that the threat of legislation may have accelerated the detergent makers' conversion to the new surfactant (ibid.). For Reuss's version of the solution to the foaming problem, see *Phosphate Hearings, December 1969,* pp. 32–33.

19. U.S. Department of Interior press release, August 4, 1967, Ohio WPCB Records, series 1800, box 19, file 3; and *Phosphate Hearings, December 1969,* pp. 129–37.

20. Anselmo Dappert, memo to WPCB Members, February 24, 1961, Rockefeller Papers (microfilm), reel 46, series 1959–1962, "Water Pollution, 1960–1961" file. The New York WPCB had delayed classification of the St. Lawrence River because of concerns that more stringent water quality requirements on the American side might encourage industry to relocate to the Canadian shore. Dappert hoped that IJC involvement would lead to roughly similar standards on both sides of the river. Dappert did not express similar fears with regard to Lakes Erie and Ontario. In these cases, he wanted to ensure that the Canadians also took significant abatement action so that New York's efforts would not be wasted.

21. Richard D. Kearney, State Department, letter to Anthony Celebrezze, HEW secretary, February 18, 1964, HEW Records, A/N 69A-1793, box 46, "Pollution, June–August 1964," file; and William R. Tyler, assistant secretary of state, letter to IJC, October 7, 1964, ibid., "January–March 1965" file.

22. "Rx for Ailing Lakes—A Low Phosphate Diet," *Environment Science and Technology* 3 (December 1969): 1245. After public hearings, the IJC issued its official report the following year, which largely reproduced the findings and recommendations of the advisory boards. See *IJC Report, 1970.*

23. For the text of Reuss's bill, see *Phosphate Hearings, December 1969,* pp. 225–33. This bill would have become an amendment to the Federal Water Pollution Control Act. The bill was referred to the House Committee on Public Works but did not advance beyond this point.

24. *Phosphate Hearings, December 1969,* pp. 2–3, 134–37. Reuss read an excerpt from the records of a recent joint task force meeting where a proposal was made to prepare a movie on the eutrophication problem with the following objectives: "To present an objective, factual explanation of eutrophication. To show that there are no quick or easy solutions to eutrophication; also to indicate that, while serious, it is not a crisis problem except in a few areas" (p. 136).

25. Ibid., pp. 70–71, 234–40.

26. Ibid., p. 84. The IJC estimated that 70 percent of the total phosphorus contributed to Lake Erie was attributable to municipal and industrial sources, while the comparable figure for Lake Ontario was 57 percent. The IJC noted that "nutrient runoff from agricultural lands is considered to be a source of pollution, but there is limited reliable information at the present time on the magnitude of this contribution" (*IJC Report, 1970,* vol. 1, pp. 9 [quote], 69–70).

27. *Phosphate Hearings, December 1969,* pp. 106, 72.

28. For Procter and Gamble's influence in the Nixon administration, see Zwick and Benstock, *Water Wasteland,* pp. 69, 75, 79–80. Many of the internal communications in the Nixon White House on this topic make reference to the views and concerns of Procter and Gamble (see Whitaker Files, box 48, "Detergents—1970" file no. 1, and box 2, "January–April 1970" file).

29. Whitaker, memo to Harlow, January 27, 1970, Whitaker Files, box 48, "Detergents—1970" file no. 1; John Ehrlichman, memo to Nixon, not dated, circa mid-March 1970, Whitaker Files, box 2, "January–April 1970" file.

30. Whitaker, memo to Russell Train, CEQ chairman, February 20, 1970, Whitaker Files, box 2, "January–April 1970" file; Train, memo to Nixon, March 17, 1970, and Whitaker, memo to Commerce secretary Maurice Stans, Interior secretary Walter Hickel, and Train, March 17, 1970, both in Whitaker Files, box 48, "Detergents—1970" file no. 2.

31. Train, memo to Nixon, March 17, 1970, and Al Alm, memo to Train, April 14, 1970, both in Whitaker Files, box 48, "Detergents—1970" file no. 2.

32. "'Curb Detergents to Save Great Lakes'," *Rochester Democrat and Chronicle,* February 5, 1970; "U.S. Agency Issues Phosphate List," *New York Times,* September 6, 1970.

33. "Russell E. Train Biography," in Train *Oral History Interview,* pp. v–vii, 1–5. See also Michael S. Durham, "Nice Guy in a Mean Job: From CEQ to EPA—A Profile of Russell Train," *Audubon* (January 1974): 97–104.

34. See Al Alm, memo to Train, April 14, 1970, Whitaker Files, box 48, "Detergents—1970" file no. 2.

35. Train, memo to Mr. Lynn, April 28, 1970, Whitaker Files, box 48, "Detergents—1970" file no. 2 (quote); and Train, memo to Nixon, May 22, 1970, Staff Member and Office Files: Edward David, Office of Science and Technology, Nixon Presidential Materials Staff, box 9, "White House, Environment, vol. III, 1971" file, National Archives. For Stans's reaction, see Ehrlichman, memo to Nixon, June 12, 1970, Whitaker Files, box 2, "May–August 1970" file.

36. "Consolidated Statistical Data Relating to Phosphates in Detergents," Whitaker Files, box 48, "Detergents-1970" file no. 1. NTA was best suited for liquid detergents. The compound absorbed moisture from the air, which caused box contents in powdered brands to solidify on the shelf. Industry executives believed that most housewives preferred detergent flakes. See "Cleaner Cleansers: Detergent Industry Assailed for Pollution, Seeks New Ingredients," *Wall Street Journal,* December 9, 1970.

37. "'Solution' to Detergent Pollution Creates More Problems as NTA Becomes Suspect," *Wall Street Journal,* July 21, 1970.

38. "Why Detergent Makers Are Turning Gray," *Business Week,* February 20, 1971, p. 66; and Maurice Stans, "Wait a Minute," in *Business and Environment: Toward Common Ground,* ed. Jeffrey H. Leonard, J. Clarence Davies, and Gordon Binder (Washington, D.C.: Conservation Foundation, 1977), p. 126. For an interesting personal account of the federal decision to oppose the use of NTA and the subsequent meeting with detergent industry officials, see Quarles, *Cleaning Up America,* pp. 28–32.

39. Testimony of Mrs. David Phillips of Housewives to End Pollution in U.S. Congress, House Committee on Government Operations, *Phosphates and Phosphate Substitutes in Detergents (Parts 1 and 2): Hearings before the Subcommittee on Conservation and Natural Resources,* 92d Cong., 1st sess., October 20, 27–29, 1971 (hereafter *Phosphate Hearings, October 1971*), pp. 465–66; and testimony of Patricia Kaltwasser of Housewives to End Pollution in *Lake Erie Conference, June 1970,* pp. 417–19. In Canada, a

number of women's groups actively lobbied for government restrictions on detergent phosphate content. See William McGucken, "The Canadian Federal Government, Cultural Eutrophication, and the Regulation of Detergent Phosphates, 1970," *Environmental Review*, 13, no. 3–4 (fall–winter 1989): 159.

40. Zwick and Benstock, *Water Wasteland*, pp. 82–83; and testimony of councilman Carl Levin of Detroit in U.S. Congress, Senate Committee on Commerce, *The Toxic Substances Control Act of 1971 and Amendment: Hearings before the Subcommittee on the Environment*, 92d Cong., 1st sess., October 1, 15, 29, November 5, 1971 (hereafter *Toxic Substances Hearings, October 1971*), p. 651. The Chicago ban would have no direct effect on the eutrophication problem in Lake Michigan since the city's effluent discharged into the Illinois River–Mississippi River systems. The ordinance was meant to help pressure the detergent industry into eliminating phosphates, which would remove them from the Lake Michigan Basin and other areas subject to accelerated eutrophication.

41. *Wall Street Journal*, April 14, 1971; "Local Phosphate Ban Legal, New York State Judge Says," *Wall Street Journal*, December 28, 1971; and "Governor Approves Pension-Cost Study," *New York Times*, June 29, 1971. For a listing of all current local, state, and federal detergent phosphate legislation as of October 1971, see the prepared statement of LeRoy H. Hurlbert, vice president of Colgate-Palmolive, in *Phosphate Hearings, October 1971*, pp. 182–90.

42. "Cleaner Cleansers: Detergent Industry, Assailed for Pollution, Seeks New Ingredients," *Wall Street Journal*, December 9, 1970; "Environment Bills Stalled by Legislative Lobbyists," *New York Times*, August 23, 1971; and testimony of Carl Levin in *Toxic Substances Hearings, October 1971*, p. 652. By October 1971, according to Colgate-Palmolive vice president LeRoy H. Hurlbert, seven states, eight counties, and fifty-six municipalities had enacted some type of phosphate legislation. Based on the provisions of these laws, by July 1, 1973, over 41 million people, or about a quarter of the American population, would be living in areas banning phosphate detergents. Of that total, approximately 29 million, or just over 70 percent, would be residents of states bordering the Great Lakes. See *Phosphate Hearings, October 1971*, pp. 180, 189–90.

43. "Detergent Makers Win a Round, but Federal Restrictions Loom," *Industry Week*, June 14, 1971, p. 28; *Chemical Week*, May 12, 1971, p. 15; and "Local Phosphate Ban Legal, New York State Judge Says," *Wall Street Journal*, December 28, 1971.

44. "New Phosphate Ban in New York State," *Wall Street Journal*, June 9, 1971.

45. "'Big Three' Choose Chicago for Phosphate Showdown," *Chemical Week*, June 7, 1972, p. 16. For a listing of some of the highest-selling nonphosphate detergents and their manufacturers, see *Phosphate Hearings, October 1971*, pp. 17–18.

46. *Phosphate Hearings, October 1971*, pp. 4–5; and Rukeyser, "Fact and Foam," pp. 167–68. Procter and Gamble continued to market detergents with phosphorus levels of up to 14.6 percent. The Canadian federal government's policy toward detergent phosphates is analyzed and contrasted with that of the U.S. federal government in McGucken, "Canadian Federal Government." For a highly critical assessment of the phosphate-free detergents, see "Ecologic Detergents: Will the Bubble Burst?" *Chemical Week*, April 28, 1971, p. 28.

47. Whitaker, memo to Ruckelshaus, February 19, 1971, Whitaker Files, box 3, "January–March, 1971" file; Rukeyser, "Fact and Foam," p. 170; and "Statement of Russell Train before Federal Trade Commission, April 26, 1971," Whitaker Files, box 48, "Detergents—1971" file no. 2, pp. 4–6.

48. "Detergents: What's a Mother to Do?" *Wall Street Journal*, October 22, 1971;

and Rukeyser, "Fact and Foam," p. 168. Most of the text of the joint statement, as well as the immediate reaction of detergent manufacturers, is contained in "Government Reversal on Phosphate Detergents," *Soap/Cosmetic/Chemical Specialties, October 1971,* pp. 33–35 (Steinfeld quote, p. 34).

49. Whitaker, memo to Ehrlichman, October 19, 1971, Whitaker Files, box 4, "July–October 1971" file; and "Ruckelshaus Disagrees on Using Phosphates," *Cleveland Press,* September 27, 1971.

50. Testimony of Griffin in *Toxic Substances Hearings, October 1971,* pp. 310–12; *Phosphate Hearings, October 1971,* pp. 180, 192.

51. *Toxic Substances Hearings, October 1971,* pp. 351–52; *Phosphate Hearings, October 1971,* pp. 159–60. During 1971, Train continued to push for legislation or a voluntary agreement that would reduce detergent phosphate levels. He wanted to immediately place a ceiling of 8.7 percent phosphorus content on all detergents, with later reductions to follow. Plans for further reductions were based on the assumption that NTA would eventually be cleared for use, but federal health authorities continued to withhold approval until further testing was done. While an 8.7 percent phosphorus ceiling received support from manufacturers that were already below this limit, some administration officials believed that such a limitation with no promise of further reductions would draw the ire of environmentalists. An 8.7 percent ceiling would still allow large quantities of detergent phosphates to enter Lake Erie and other endangered waters. (In the state and local laws regulating detergent phosphate content, the 8.7 percent limit was only a temporary stage before the complete ban went into effect.) See a number of memos relating to this subject in Whitaker Files, box 48, "Detergents—1971" file no. 2.

52. *Phosphate Hearings, October 1971,* pp. 4–5 (Chicago mayor Richard Daley expressed similar sentiments; see *Toxic Substances Hearings, October 1971,* pp. 368–71), and pp. 469–70.

53. For the FDA test results and accompanying information, see *Phosphate Hearings, October 1971,* pp. 8–18. See the testimony of officials representing Procter and Gamble, Lever Brothers, and Colgate-Palmolive in ibid.

54. Ibid., p. 467 (quote); *Toxic Substances Hearings, October 1971,* pp. 340–48.

55. See the testimony of officials representing Procter and Gamble, Lever Brothers, and Colgate-Palmolive in *Phosphate Hearings, October 1971;* J. R. Vallentyne, "Phosphorus and the Control of Eutrophication," *Canadian Research and Development* (May–June 1970): 36–49; and McGucken, "Canadian Federal Government," p. 163.

56. "'Big Three' Choose Chicago for Phosphate Showdown," *Chemical Week,* June 7, 1972, pp. 16–17; and Milliken, letter to George C. Edwards, city clerk of Detroit, January 24, 1972, Michigan State Archives, RG 78-47, box 1, folder 7. Milliken also believed that, in general, regulations of this nature should be made at the state or national level. "In Matter of Proposed Regulation Banning Phosphates in Detergents . . .," March 14, 1972, in Illinois Pollution Control Board, *Opinions,* vol. 4, pp. 71–92, Illinois State Library, Springfield. The board also cited the lack of a eutrophication problem in most state waters.

57. See McGucken, "Canadian Federal Government."

58. See the letter from Canadian ambassador Marcel Cadieux to Representative Reuss, October 14, 1971, printed in *Phosphate Hearings, October 1971,* pp. 20–21; and Don Munton, "Dependence and Interdependence in Transboundary Environmental Relations," *International Journal 36,* no. 1 (winter 1980–1981): 152–57.

59. Reuss is quoted in David Zwick, "Water Pollution," in *Nixon and the Environ-*

ment: The Politics of Devastation, ed. James Rathlesberger (New York: Taurus Communications, 1972), p. 50. For similar comments by Senator Muskie, see "Muskie Scores Nixon on Lake Compact," *New York Times,* April 9, 1972.

60. "Local Phosphate Ban Legal, New York State Judge Says," *Wall Street Journal,* December 28, 1971.

61. "'Big Three' Choose Chicago," p. 16; "Phosphate Washout," *Chemical Week,* June 14, 1972, p. 16;

62. "'Big Three' Choose Chicago," p. 17; *Chemical Week,* July 12, 1972, p. 33; and "City's Phosphate Ban Upheld," *Chicago Tribune,* May 20, 1975, p. 3.

63. Procter and Gamble press release, October 10, 1972, Whitaker Files, box 48, "Detergents—1971" file no. 1. In May 1972, Dr. Merlin K. DuVal, assistant secretary for Health and Scientific Affairs in the HEW, announced that the department would continue to oppose the use of NTA in laundry detergents. DuVal cited continuing concerns about the possible carcinogenic and mutagenic effects of NTA. See HEW press release, May 5, 1972, in the above file.

64. "Are Phosphate Detergents on the Shelf to Stay?" *Chemical Week,* March 21, 1973, p. 28; and "Diamond Backs June 1 Deadline on Ban of Detergent Phosphates," *New York Times,* February 14, 1973.

65. See the testimony of a number of participants in *National Conference on Water Pollution, 1960;* J. Samuel Walker, "Nuclear Power and the Environment: The Atomic Energy Commission and Thermal Pollution, 1965–1971," *Technology and Culture* 30, no. 4 (October 1989): 970–71.

66. Martin Melosi, *Coping with Abundance: Energy and Environment in Industrial America* (Philadelphia: Temple University Press, 1985), p. 200; and Charles Komanoff, Holly Miller, and Sandy Noyes, *The Price of Power: Electric Utilities and the Environment* (New York: Council on Economic Priorities, 1972), p. 48.

67. Elizabeth S. Rolph, *Nuclear Power and the Public Safety* (Lexington, Mass.: Lexington Books, 1979), p. 79; and Melosi, *Coping with Abundance,* pp. 233, 239. But as of September 1968, only fourteen nuclear plants, generating about 2,800 megawatts of power, were in operation. Seaborg quoted from Walker, "Nuclear Power and the Environment," pp. 965–68 (968).

68. Theodore L. Brown, *Energy and the Environment* (Columbus, Ohio: Charles E. Merrill Publishing, 1971), pp. 72–74. In chapter 7, Brown provides a clearly written technical overview of power plant thermal pollution and the means of controlling it.

69. Komanoff et al., *Price of Power,* pp. 55–60; and Brown, *Energy and the Environment,* pp. 76–82. Cost estimates for the installation and operation of closed-cycle cooling systems varied greatly, depending on the source and the particular application in question. These figures are taken from "Utilities Burn over Cooling Towers," *Business Week,* April 3, 1971, p. 52.

70. The cooling tower system just described was referred to as a "wet-tower." There was also a "dry-tower" system. In this method, employed in water-scarce areas, water was recirculated through a series of closed tubings in a system similar to that of an automobile radiator. This system eliminated evaporative water loss, but high capital and operating costs, as well as some other limitations, reduced the attractiveness of dry towers, except in special circumstances.

71. Udall is quoted from U.S. Congress, Senate Committee on Public Works, *Water Pollution—1967 (Part 2): Hearings before the Subcommittee on Air and Water Pollution,* 90th Cong., 1st sess., August 9–10, 1967, p. 539. See also Donald F. Hornig, special

assistant to the president for Science and Technology, memo to Joseph Califano, December 21, 1967, WHCF, Office Files of James Gaither, box 39, "Water Pollution" file, Lyndon Baines Johnson Presidential Library.

72. Melosi, *Coping with Abundance,* pp. 230–39; Walker, "Nuclear Power and the Environment," pp. 975–79.

73. *Proceedings, Second Session of the Conference in the Matter of Pollution of Lake Michigan and Its Tributary Basins, Chicago, February 25, 1969* (Washington, D.C.: FWPCA, 1969) (hereafter *Lake Michigan Conference, February 1969*), pp. 543–44; "Thermal Ban Set for Lake Michigan," *New York Times,* September 21, 1970.

The only nuclear plant in operation at this time was Big Rock Point (Michigan), a small pilot plant with limited generating capacity operated by Consumers Power Company. The other six plants were Northern Indiana Public Service Company's Bailly plant (Ind.), Commonwealth Edison Company's Zion plant (Ill.), Indiana and Michigan Electric Company's Cook plant (Mich.), Consumers Power Company's Palisades plant (Mich.), Wisconsin Electric Power Company's Point Beach plant (Wis.), and the Kewaunee plant (Wis.), which was jointly owned by Wisconsin Public Service Corporation, Wisconsin Power and Light Company, and Madison Gas and Electric Company.

74. For Udall's comments, see *Lake Michigan Conference, January–March 1968,* p. 110.

75. *Lake Michigan Conference, February 1969,* pp. 541–46.

76. Ibid., pp. 555–57, 621–24, 546, 687–89. The size of the "mixing zone" (an area around the discharge point where standards were not applied) would be an important point of contention in the ensuing controversy over thermal discharge control. The most ardent environmentalists wanted to completely eliminate the use of mixing zones.

The methods for effective treatment and disposal of liquid radioactive waste were well known, and the committee expressed less concern about this issue. A major legal controversy did arise, however, when the Minnesota Pollution Control Agency attempted to impose its own standards for gaseous and liquid radioactive wastes on the new Monticello nuclear plant. The federal courts eventually ruled that the AEC should retain sole authority for setting and enforcing such standards. See Richard S. Lewis, *The Nuclear-Power Rebellion: Citizens vs. the Atomic Industrial Establishment* (New York: Viking Press, 1972), pp. 122–35.

77. "The Dilemma in Pollution Fighting," *Wall Street Journal,* November 10, 1970.

78. "A New River," *Environment* 12 (January–February 1970): 37; and Walker, "Nuclear Power and the Environment," pp. 987–88. Fossil fuel power plants that were at least partially constructed on navigable waters required a permit from the Army Corps of Engineers. Thus, under the new law the majority of fossil fuel plants would also be subject to state clearance.

79. *Lake Michigan Conference, March–April 1970,* p. 669; *Lake Michigan Conference, May 1970,* p. 15.

80. *Lake Michigan Conference, May 1970,* pp. 122–36; "U.S. Lays Down Tough Standard on Lake Heat Pollution," *Chicago Sun-Times,* May 8, 1970. See also John G. Quale, president of Wisconsin Electric, letter to Governor Knowles, May 8, 1970, and James R. Underkofler, president of Wisconsin Power and Light, letter to Knowles, May 18, 1970, Knowles Papers, series 2142, box 36, file 39.

81. "Hickel Aide Quits Pollution Post," *New York Times,* September 19, 1970; "Study Panel to Seek Safe Heat Levels for Lake Michigan," *Chicago Tribune,* October 30, 1970.

82. *Proceedings, Third Session Reconvened in Workshop Sessions of the Conference in the Matter of Pollution of Lake Michigan and Its Tributary Basin, Chicago, September*

28–30 and October 1–2, 1970 (Washington, D.C.: EPA, 1971). The first four days were devoted to technical presentations, the last day heard general statements from the public. See also "The Dilemma in Pollution Fighting," *Wall Street Journal,* November 10, 1970.

83. "Federal Officials Devise Plan to Set 'Heat Quotas' on Industrial Discharges into Waterways," *New York Times,* November 1, 1970.

84. Ibid.; and Dingell press release, October 30, 1970, EPA Records, A/N 74-22, box 52, "Lake Michigan (Four-State), Clips" file.

85. Rolph, *Nuclear Power and the Public Safety,* p. 37; and Steve Ebbin and Ronald Kasper, *Citizen Groups and the Nuclear Power Industry: Uses of Scientific and Technological Information* (Cambridge, Mass.: MIT Press, 1974), pp. 139–41. Ebbin and Kasper discuss the entire AEC licensing process in great detail on pp. 33–56, 139–62.

86. Melosi, *Coping with Abundance,* pp. 234–39.

87. Ebbin and Kasper, *Citizen Groups,* pp. 9–14, 190–93. The literature on the antinuclear movement is very large. See Jerry W. Mansfield, *The Nuclear Power Debate: A Guide to the Literature* (New York: Garland Publishing, 1984); and Jerome Price, *The Antinuclear Movement* (Boston: Twayne Publishers, 1982). For two scholarly appraisals of the federal government's response to the movement, see J. Samuel Walker, *Containing the Atom: Nuclear Regulation in a Changing Environment, 1963–1971* (Berkeley and Los Angeles: University of California Press, 1992); and Brian Balogh, *Chain Reaction: Expert Debate and Public Participation in American Commercial Nuclear Power, 1945–1975* (Cambridge, England: Cambridge University Press, 1991).

88. Frances Gendlin, "The Palisades Protest: A Pattern of Citizen Intervention," *Bulletin of the Atomic Scientists* 27 (November 1971): 53–55; and Lewis, *Nuclear-Power Rebellion,* pp. 135–41.

89. Gendlin, "Palisades Protest," p. 55; and Lewis, *Nuclear-Power Rebellion,* p. 141.

90. Gendlin, "Palisades Protest," pp. 53–56; Lewis, *Nuclear-Power Rebellion,* pp. 136–41; and *Nucleonics Week,* March 18, 1971, pp. 3–4. Under the agreement, Consumers Power had forty-two months to install a natural draft cooling tower or thirty-two months to complete a mechanical draft tower.

91. *Nucleonics Week,* March 25, 1971, pp. 4–5 (5).

92. Ibid. Ironically, BPI and other environmental groups intervened in the construction license hearing for the Bailly plant anyway, on the grounds that the 450-foot cooling tower would violate the aesthetic appeal of the nearby Indiana Dunes National Park. David Comey, environmental director for BPI, argued that since the plant would be utilizing closed-cycle cooling it could be moved farther inland.

93. *Lake Erie Conference, June 1970,* pp. 159–61.

94. Toledo Edison Company press release, July 31, 1970, Rhodes Papers, series 353, box 22, folder 14. The company estimated that it would cost $9 million to build the tower and $3 million annually to operate. For Detroit Edison, see *Nucleonics Week,* May 13, 1971, pp. 7–8.

95. Walker, "Nuclear Power and the Environment," p. 989; and Rolph, *Nuclear Power and the Public Safety,* pp. 104–5.

96. *Lake Michigan Conference, March 1971,* pp. 10–15. Shortly before the opening of the conference, EPA Region 5 administrator Francis Mayo recommended the adoption of this policy to David Dominick, acting commissioner of the EPA Water Quality Office. Dominick then made the same recommendation to Ruckelshaus. See Mayo, memo to Dominick, March 12, 1971, and Dominick, memo to Ruckelshaus, March 18, 1971, EPA Records, A/N 74-22, box 50, "Lake Michigan (Four-State), Admin." file no. 2.

97. *Lake Michigan Conference, March 1971,* pp. 52–57.

98. Ibid., pp. 70–74, 139–41.

99. Ibid., pp. 573–78; and *Nucleonics Week,* March 11, 1971, pp. 5–6. Officials from the AEC and the Federal Power Commission also appeared and offered testimony that supported the position of the power industry.

100. *Lake Michigan Conference, March 1971,* pp. 175–83.

101. Ibid., pp. 767–90; and "Split over Lake A-Plan," *Chicago Tribune,* March 26, 1971.

102. Illinois Pollution Control Board, "In the Matter of Thermal Standards, Lake Michigan," June 9, 1971, Division of Environmental Protection Records, A/N 1989-257, box 4, "Lake Michigan Enforcement Conference, March 1971" file, Wisconsin State Archives.

103. For Michigan, see MWRC Minutes, August 19–20, 1971, pp. 5–6; and *Nucleonics Week,* September 16, 1971, p. 3. At the Palisades plant Consumers Power was still bound by its agreement with the intervenors to install closed-cycle cooling. For Wisconsin, see "Lake Michigan Thermal Standards," following Wisconsin Natural Resources Board Minutes, December 8, 1971, Wisconsin DNR, Madison; and "State Softens Rules on Thermal Pollution," *Madison Capital Times,* December 9, 1971.

104. Mayo, memo to David Dominick, March 29, 1971, EPA Records, A/N 74-22, box 50, "Lake Michigan (Four-State), Admin." file no. 2; Mayo, letter to Colonel Metullus A. Barnes Jr., district engineer, Army Corps of Engineers (Chicago), July 23, 1971, and Mayo, letter to Colonel Myron D. Snoke, district engineer, Army Corps of Engineers (Detroit), July 23, 1971, EPA Records, A/N 75-63, box 2, "Lake Michigan (Four-State) Legal," file no. 3.

105. Quarles, memo to Ruckelshaus, March 10, 1972, EPA Records, A/N 74-30, box 3, "Region 5" file no. 1. In addition, if the case ended up in court, Quarles believed there was a good chance the judge would rule against the EPA on the grounds that the agency had no authority under statute to impose effluent standards or to deal with future—not existing—pollution.

106. *Nucleonics Week,* July 6, 1972, pp. 6–7; Rolph, *Nuclear Power and the Public Safety,* pp. 101–17.

107. Rolph, *Nuclear Power and the Public Safety,* pp. 132–33. For a detailed account of the Calvert Cliffs case, see Lewis, *Nuclear-Power Rebellion,* pp. 269–87.

Epilogue: The 1970s and Beyond

1. *IJC Report, 1970,* vol. 1; and Munton, "Transboundary Environmental Relations," pp. 151–52.

2. Fitzhugh Green, memo to Ruckelshaus, November 2, 1971, and William Mansfield, memo to Green, November 15, 1971, EPA Records, A/N 75-56, box 2, "Canada" file.

3. U.S. Department of State, "Agreement between the United States of America and Canada on Great Lakes Water Quality," April 15, 1972, TIAS no. 7312, in *United States Treaties and Other International Agreements,* vol. 24, pt. 1.

4. For analysis of the agreement's provisions, see Munton, "Transboundary Environmental Relations," p. 159; and John E. Carrol, *Environmental Diplomacy: An Examination and a Prospective of Canadian-U.S. Transboundary Environmental Relations* (Ann Arbor: University of Michigan Press, 1983), pp. 131–32.

5. Quote is from Walter Rosenbaum, *Environmental Politics and Policy,* 3d ed.

(Washington, D.C.: Congressional Quarterly Press, 1995), pp. 215–17 (217); Harvey Lieber, *Federalism and Clean Waters: The 1972 Water Pollution Control Act* (Lexington, Mass.: Lexington Books, D. C. Heath and Company, 1975), pp. 7–8.

6. See, for example, the testimony of representatives from the American Iron and Steel Institute, the American Petroleum Institute, the Manufacturing Chemists Association, and the American Paper Institute in U.S. Congress, Senate Committee on Public Works, *Water Pollution Control Legislation (Part 2): Hearings before the Subcommittee on Air and Water Pollution,* 92d Cong., 1st sess., March 1971.

7. Diamond, letter to Milliken, September 17, 1971, Milliken Papers, box 757, "Water Pollution" file, p. 2. With a few exceptions, the state governors who provided testimony and submitted statements on the proposed legislation paid little attention to administrative provisions, instead concentrating on the size of the grant package.

8. Lieber, *Federalism and Clean Waters,* p. 54; and Zwick and Benstock, *Water Wasteland.*

9. Whitaker, *Striking a Balance,* pp. 79–88; and Quarles, *Cleaning Up America,* ch. 8.

10. U.S. Council on Environmental Quality, *Environmental Quality: Fourth Annual Report* (Washington, D.C.: GPO, 1973), pp. 171–72.

11. Rosenbaum, *Environmental Politics and Policy,* 1st ed. (Washington, D.C.: Congressional Quarterly Press, 1985), pp. 153–55.

12. Ibid.

13. Whitaker, *Striking a Balance,* pp. 84–90. In view of the repeated failure to appropriate all the money authorized for sewage treatment plant construction under previous laws, the authors of the 1972 amendments to the Federal Water Pollution Control Act placed the yearly construction grant authorizations under contract obligation authority, which meant that these outlays would bypass the normal budgetary and appropriations process.

14. Kurt M. Hunciker, "The Clean Water Act and Related Developments in the Federal Water Pollution Control Program during 1977 (Part 2): Modifications of the Municipal Program," *Harvard Environmental Law Review* 2 (1977): 130–31; and Lieber, *Federalism and Clean Waters,* pp. 111–19.

15. "Great Lakes Pollution Cleanup Stagnates as Problems Mount," *Engineering News-Record,* May 27, 1976, pp. 26–27; and "Waste Woes: Municipalities Trail Industry in Cleanup of Water Pollution," *Wall Street Journal,* October 13, 1976.

16. Marcus, *Promise and Performance,* pp. 150–57.

17. Ibid.; and U.S. Council on Environmental Quality, *Environmental Quality: Fifth Annual Report* (Washington, D.C.: GPO, 1974), pp. 141–42.

18. "EPA's Effluent Guidelines Have Their Day in Court," *Chemical Week,* October 30, 1974, pp. 33, 36; and "Industry and the Permit System," *Journal, WPCF* 45 (November 1973): 2264–65.

19. See testimony of John Quarles, deputy administrator of the EPA, in U.S. Congress, House Committee on Public Works, *Implementation of the Federal Water Pollution Control Act: Hearings before the Subcommittee on Investigations and Review,* 93d Cong., 2d sess., February–June 1974, pp. 478–79, 493. Lettie M. Wenner, *The Environmental Decade in Court* (Bloomington: Indiana University Press, 1982), pp. 49–52; and "Effluent Rules: Progress out of Court," *Chemical Week,* November 19, 1975, p. 39.

20. U.S. Council on Environmental Quality, *Fifth Annual Report,* pp. 141–42; and Marcus, *Promise and Performance,* pp. 152, 156–58.

21. Great Lakes Water Quality Board, *Great Lakes Water Quality: Fourth Annual*

Report to the International Joint Commission (1976), p. 75; and "Great Lakes Pollution Cleanup Stagnates as Problems Mount," *Engineering News-Record,* May 27, 1976, pp. 26–27. For the national scene, see the testimony of EPA administrator Russell Train in U.S. Congress, House Committee on Public Works and Transportation, *Implementation of the Federal Water Pollution Control Act: Hearings before the Subcommittee on Investigations and Review,* 94th Cong., 2d sess., February 1976, pp. 11–12. According to Train, approximately five hundred major industrial dischargers had contested EPA-issued permits. Quarles and other EPA officials usually placed the number of total major industrial dischargers at between 2,700 and 3,000.

22. U.S. Council on Environmental Quality, *Environmental Quality: Sixth Annual Report* (Washington, D.C.: GPO, 1975), p. 66.

23. "EPA Promises Tough Enforcement of New Water Pollution Law," *Chemical and Engineering News,* November 6, 1972, p. 17; and U.S. Council on Environmental Quality, *Sixth Annual Report,* pp. 66–68.

24. The following summary is drawn from Bastow, *"This Vast Pollution . . . ,"* and Bartlett, *The Reserve Mining Controversy.*

25. Company officials placed the total cost of the new facilities at $370 million, but the IRS later ruled that approximately $130 million of that total was for product improvement and process changes. See Bartlett, *The Reserve Mining Controversy,* p. 206. In May 1976, federal district court judge Edward Devitt fined Reserve $837,500 for violation of the company's state effluent discharge permit and an additional $200,000 for improperly withholding evidence during the federal trial. See Bastow, *"This Vast Pollution . . . ,"* pp. 180–83.

26. "Scott's Pollution Fighter Girding for New Battles," *Chicago Tribune,* February 1, 1976.

27. Victoria Lillie, "Compliance Monitoring: Where the Action Is in Water Enforcement," *Environment Midwest* (May 1975): 4–5, 16; and J. T. Sliter, "Enforcement of Municipal Permits: Portent of Problems," *Journal, WPCF* 48 (March 1976): 422–24.

28. Mayo, memo to Train, July 2, 1974, and Indiana State Board of Health, "Preliminary Report of Phosphorus Trends at Municipal Sewage Treatment Plants and in Indiana Streams for Years 1971, 1972, and 1973," EPA Records, A/N 76-35, box 7, "Region 5" file. The report found (p. 6) that there had been, on average, "a 60% reduction in the phosphorus concentration in raw sewage and a corresponding decrease in the phosphorus concentration of final sewage treatment plant effluent."

29. See, for example, IJC, *Fifth Annual Report, Great Lakes Water Quality* (for 1976), sections 1 and 7.

30. Train, memo to Alexander, December 3, 1976, EPA Records, A/N 78-21, box 2, "Region 5" file no. 4.

31. "Statement of George Alexander before MWRC, August 27, 1976," pp. 4–5, EPA Records, A/N 78-21, box 2, "Region 5" file no. 3. In 1977, the EPA backed an amendment introduced by Senator Gaylord Nelson that would have effectively banned the use of laundry detergents containing phosphates in the Great Lakes states. The amendment passed the Senate by a vote of 77 to 17, but members of the House rejected the measure in conference prior to the passage of the Clean Water Act of 1977 without offering an alternative. See "Senate Votes Phosphate Ban," *Chemical Week,* August 17, 1977, p. 20; and *Environment Reporter,* November 7, 1977, p. 1027.

32. "Minnesota Will Ban Phosphates," *Milwaukee Journal,* April 30, 1976.

33. "Phosphate Ban Is Sought," *Detroit News,* September 24, 1976; and MWRC

Minutes, January 20–21, 1971, p. 4. According to the MWRC, Detroit was removing only 50 percent of the city's sewage phosphorus content by 1976. See "Laundry Phosphate Assailed," *Detroit News,* August 26, 1976.

34. MWRC Minutes, February 24–25, 1971, p. 3; "Curb on Phosphates Begins in Michigan," *New York Times,* October 2, 1977; and "Wisconsin Limits Detergent Phosphates," *Environment Midwest* (July 1978): 12. Each of the Great Lakes states with bans limited detergent phosphate content to 0.5 percent phosphorus. Despite objections from the IJC, Canada allowed a phosphorus content of up to 2.2 percent phosphorus.

35. *Nucleonics Week,* November 16, 1972, pp. 3–4.

36. Rolph, *Nuclear Power and the Public Safety,* pp. 137–39.

37. "EPA Sets Easier Water-Discharge Rules for Steam Plants than It First Proposed," *Wall Street Journal,* October 3, 1974; and *Environment Reporter,* October 4, 1974, p. 841.

38. U.S. Council on Environmental Quality, *Environmental Quality: Ninth Annual Report* (Washington, D.C.: GPO, 1978), pp. 108–10; and IJC, *Sixth Annual Report on Great Lakes Water Quality* (Washington and Ottawa, 1979), pp. 10, 13.

39. P. J. Piecuch, "EPA Charts Policy for Deadline Enforcement," *Journal,* WPCF 49 (July 1977): 1569–71; Hunciker, "The Clean Water Act," pp. 133–34.

40. Great Lakes Water Quality Board, *Great Lakes Water Quality: Sixth Annual Report to the International Joint Commission* (July 1978), pp. 21–28.

41. "U.S. Asks Ohio Action on Twenty Polluting Firms," *Cleveland Plain Dealer,* March 28, 1978. Williams quoted from "U.S. Too Hasty on Pollution, State Says," *Cleveland Plain Dealer,* March 30, 1978. "Michigan DNR Special Task Force Report on Enforcement," December 28, 1977, Milliken Papers, box 606, "Bill Rustem" file. In a 1982 article, two scholars concluded after reviewing the literature on state air and water pollution enforcement efforts—mainly from the 1970s—that industry-government bargaining continued to be the primary determinant of treatment requirements and the timetables for implementing these controls. The literature also revealed that agency inspections were infrequent and that informal negotiation, rather than formal legal action, remained characteristic of the enforcement process. See Paul B. Downing and James N. Kimball, "Enforcing Pollution Control Laws in the U.S.," *Policy Studies Journal* 11, no. 1 (September 1982): 55–65.

42. For business opposition to environmental policy in the 1970s, see Hays, *Beauty, Health, and Permanence,* pp. 307–28. For the resurgence of business executives' political influence in the 1970s, see Vogel, *Fluctuating Fortunes,* chs. 6–7.

43. See, for example, the comments of officials from U.S. Steel and Jones and Laughlin Steel in Victoria Lillie, "Earth Day—Five Years Later," *Environment Midwest* (April 1975): 4–5.

44. Dunlap, "Trends in Public Opinion," pp. 96–102. Stein is quoted in Lillie, "Earth Day—Five Years Later," p. 5.

45. James Voytko, "The Clean Water Act and Related Developments in the Federal Water Pollution Control Program during 1977 (Part 1): Industrial Dischargers," *Harvard Environmental Law Review* 2 (1977): 121–25.

46. Hunciker, "The Clean Water Act," pp. 126–35.

47. Train is quoted in Jack McWethy, "The Great Lakes Get a New Lease on Life," *U.S. News and World Report,* September 27, 1976, p. 51.

48. William K. Stevens, "Great Lakes Pollution Fight Is Gaining," *Environment Midwest* (June 1974): 2–4; "Detroit River Success Story," *Detroit News,* January 23,

1976; IJC, *Second Annual Report on Great Lakes Water Quality* (Washington and Ottawa, 1974), p. 1; and "The Beach Scene," *Environment Midwest* (August 1978): 2.

49. Wisconsin DNR, *Annual Water Quality Report to Congress, 1975,* section 2, p. 63; IJC, *Seventh Annual Report on Great Lakes Water Quality* (Washington and Ottawa, 1980), pp. 28–38.

50. Theodora Colborn et al., *Great Lakes, Great Legacy?* (Washington, D.C.: Conservation Foundation, 1990), pp. 87–99.

51. IJC, *Fifth Annual Report on Great Lakes Water Quality,* section 4.

52. "Progress, not Victory, on Great Lakes Pollution," *New York Times,* May 7, 1994. The health of the Great Lakes can be monitored through the biennial reports issued by the IJC under the provisions of the 1978 Water Quality Agreement.

53. IJC, *Eighth Biennial Report under the Great Lakes Water Quality Agreement of 1978* (Washington and Ottawa, 1996), ch. 3.

54. Colborn et al., *Great Lakes, Great Legacy?* pp. 38–47.

55. Ibid., pp. 78–84, 113–30.

56. For a judicious appraisal of the flaws in the current environmental protection system from a longtime observer, see Rosenbaum, *Environmental Politics and Policy,* 3d ed.

57. This is a major theme of Hays, *Beauty, Health, and Permanence.*

58. Vogel, "The Public-Interest Movement," p. 609.

59. Ellis W. Hawley, "Social Policy and the Liberal State in Twentieth-Century America," in Federal Social Policy: The Historical Dimension, ed. Donald T. Critchlow and Hawley (State College: Pennsylvania State University Press, 1988), pp. 124–25.

Selected Bibliography

Unpublished Collections

B. F. Hamilton Library, Franklin College (Indiana)
    1. Roger Branigin Papers

Bentley Historical Library, University of Michigan
    1. John Swainson Papers
    2. George Romney Papers
    3. William Milliken Papers

Illinois State Historical Library, Springfield
    1. Otto Kerner Papers

Indiana Department of Environmental Management, Indianapolis
    1. Indiana Stream Pollution Control Board (SPCB) Minutes and Annual Reports

Indiana State Archives, Indianapolis
    1. Indiana Department of Health, Sanitary Engineering Division
    2. Matthew Welsh Papers
    3. Edgar Whitcomb Papers

Lyndon B. Johnson Presidential Library, Austin, Texas
    1. Stewart Udall Oral History Interview
    2. White House Central Files

John F. Kennedy Presidential Library, Boston, Massachusetts
    1. James Quigley Oral History Interview

Michigan Department of Natural Resources, Lansing
    1. Michigan Water Resources Commission Minutes

Michigan State Archives, Lansing
    1. Michigan Department of Natural Resources

National Archives, Washington, D.C.
    1. Federal Water Pollution Control Administration, RG 382
    2. U.S. Bureau of the Budget, RG 51

New York Department of Environmental Conservation, Albany
    1. New York Water Resources Commission Minutes

New York State Archives, Albany
    1. New York Department of Environmental Conservation
    2. Nelson Rockefeller Papers (microfilm)

Nixon Presidential Materials Staff, National Archives and Records Administration II, College Park, Maryland
    1. Staff Member and Office Files: John Whitaker

Ohio State Archives, Columbus
    1. Michael DiSalle Papers
    2. John Gilligan Papers
    3. Izaak Walton League, Ohio Division
    4. Ohio Department of Health, Water Pollution Control Board
    5. Ohio Environmental Protection Agency
    6. James Rhodes Papers

Washington National Records Center, Suitland, Maryland (some of these records may have been moved to the new National Archives facility in College Park, Md., since the time of author's research)
    1. U.S. Environmental Protection Agency, RG 412
    2. U.S. Department of Health, Education and Welfare, RG 235
    3. U.S. Public Health Service, RG 90
    4. Federal Water Pollution Control Administration, RG 382

Wisconsin Department of Natural Resources, Madison
    1. Wisconsin Natural Resources Board Minutes

Wisconsin State Archives, Madison
    1. Warren Knowles Papers
    2. Patrick Lucey Papers
    3. Gaylord Nelson Papers
    4. Wisconsin Committee on Water Pollution
    5. Wisconsin Department of Natural Resources

Published Oral Histories

1. William D. Ruckelshaus, EPA *Oral History Interview* No. 1, January 1993.
2. Russell E. Train, EPA *Oral History Interview* No. 2, July 1993.

Newspapers

Buffalo Evening News
Chicago American
Chicago Sun-Times
Chicago Tribune
Christian Science Monitor
Cleveland Plain Dealer
Cleveland Press
Detroit Free Press
Detroit News
Gary Post-Tribune
Milwaukee Journal
Milwaukee Sentinel
New York Times
Rochester Democrat and Chronicle
Wall Street Journal

Journals

American Journal of Public Health
Bulletin of the Atomic Scientists
Business Week
Chemical and Engineering News
Chemical Week
Civil Engineering
Clean Waters for Ohio
Engineering News-Record
Environment Midwest
Environment Reporter
Environmental Health Letter
Environmental Science and Technology
Field and Stream
Harvard Environmental Law Review
Industrial Water and Wastes
Industry Week
Journal, American Water Works Association
Journal, Water Pollution Control Federation
Limnology and Oceanography
National Journal
News in Engineering
Proceedings of the Annual Industrial Waste Conference, Lafayette, Indiana
Pulp and Paper
Sewage and Industrial Wastes
Soap/Cosmetic/Chemical Specialties
Wastes Engineering
Water and Sewage Works

Government Documents (in chronological order)

International Joint Commission. *Final Report of the International Joint Commission on the Pollution of Boundary Waters Reference.* Washington, D.C.: GPO, 1918.

U.S. Congress. House Committee on Public Works. *Water Pollution Control.* 80th Cong., 1st sess., June 11–13, 16, 1947.

International Joint Commission. *Report on the Pollution of Boundary Waters.* Washington, D.C., and Ottawa, 1951.

U.S. Congress. Senate Committee on Public Works. *Water and Pollution Control: Hearings before the Subcommittee on Flood Control—Rivers and Harbors.* 84th Cong., 1st sess., April 22, 25, 26, 1955.

U.S. Congress. House Committee on Public Works. *Water Pollution Control Act: Hearings before the Subcommittee on Rivers and Harbors.* 84th Cong., 1st and 2d sess., July 20, 1955, and March 12–15, 1956.

U.S. Congress. House Committee on Public Works. *Amend Federal Water Pollution Control Act: Hearings before the Subcommittee on Rivers and Harbors.* 85th Cong., 2d sess., May 20–22, 1958.

U.S. Congress. Senate Select Committee on National Water Resources. *Water Resources (Part 7).* 86th Cong., 1st sess., October 29, 1959.

Proceedings of the National Conference on Water Pollution in Washington, D.C., December 12–14, 1960. Washington, D.C.: GPO, 1961.

Proceedings of the Joint Federal–State of Michigan Conference on Pollution of Navigable Waters of the Detroit River and Lake Erie, and Their Tributaries within the State of Michigan, 1st sess., Detroit, March 27–28, 1962. Washington, D.C.: HEW, 1962.

Proceedings of the State and Interstate Water Pollution Control Administrators in Joint Meeting with the Conference of State Sanitary Engineers in Washington, D.C., May 21, 1962. Washington, D.C.: HEW, PHS, DWSPC, 1962.

U.S. Congress. House Committee on Government Operations. *Water Pollution Control and Abatement (Parts 1A and 1B—National Survey): Hearings before the Subcommittee on Natural Resources and Power.* 88th Cong., 1st sess., May–June 1963.

U.S. Congress. Senate Committee on Public Works. *Water Pollution Control: Hearings before the Special Subcommittee on Air and Water Pollution.* 88th Cong., 1st sess., June 17–20, 25–26, 1963.

U.S. Congress. House Committee on Government Operations. *Water Pollution Control and Abatement (Part 3—Chicago Area and Lower Lake Michigan): Hearings before the Subcommittee on Natural Resources and Power.* 88th Cong., 1st sess., September 6, 1963.

Proceedings of the Conference in the Matter of the Pollution of the Interstate Waters of the Menominee River and Its Tributaries, Menominee, Michigan, November 6–8, 1963. Washington, D.C.: HEW, 1964.

U.S. Congress. House Committee on Public Works. *Water Pollution Control Act Amendments.* 88th Cong., 1st and 2d sess., December 1963 and February 1964.

U.S. Congress. House Committee on Public Works. *Water Pollution Control Hearings on Water Quality Act of 1965.* 89th Cong., 1st sess., February 18, 19, 23, 1965.

Proceedings of the Conference in the Matter of Pollution of the Interstate Waters of the Grand Calumet River, Little Calumet River, Calumet River, Wolf Lake, Lake Michigan, and Their Tributaries, Chicago, March 2–9, 1965. Washington, D.C.: HEW, 1965.

Report on Pollution of the Detroit River, Michigan Waters of Lake Erie, and Their Tribu-

taries: Summary, Conclusions, and Recommendations, April 1965. Washington, D.C.: HEW, 1965.

"Proceedings of the Governors Conference on Great Lakes Pollution, in Cleveland, May 10, 1965." State Library, Columbus, Ohio.

U.S. Congress. Senate Committee on Public Works. *Water Pollution (Part 1): Hearings before the Special Subcommittee on Air and Water Pollution.* 89th Cong., 1st sess., May 19–21, 1965.

Proceedings of the Conference in the Matter of Pollution of the Navigable Waters of the Detroit River and Lake Erie, and Their Tributaries in the State of Michigan, Second Session, Detroit, June 15–18, 1965. Washington, D.C.: FWPCA, 1965.

U.S. Congress. Senate Committee on Public Works. *Water Pollution (Part 2): Field Hearings before the Special Subcommittee on Air and Water Pollution.* 89th Cong., 1st sess., June 17, 1965, pp. 789–909 (Buffalo session).

Report on Pollution of Lake Erie and Its Tributaries, July 1965. Washington, D.C.: HEW, 1965.

Proceedings of the Conference in the Matter of Pollution of Lake Erie and Its Tributaries, First Session, Cleveland, August 3–6, 1965. Washington, D.C.: FWPCA, 1965.

Proceedings of the Conference in the Matter of Pollution of Lake Erie and Its Tributaries, First Session, Buffalo, August 10–11, 1965. Washington, D.C.: FWPCA, 1965.

Proceedings of the Conference in the Matter of Pollution of the Interstate Waters of the Grand Calumet River, Little Calumet River, Calumet River, Wolf Lake, Lake Michigan and Their Tributaries, Chicago, Technical Session, January 4–5, 1966. Washington, D.C.: FWPCA, 1966.

Proceedings of the Conference in the Matter of Pollution of the Interstate Waters of the Grand Calumet River, Little Calumet River, Calumet River, Wolf Lake, Lake Michigan, and Their Tributaries, Chicago, Technical Session, February 2, 1966. Washington, D.C.: FWPCA, 1966.

U.S. Congress. Senate Committee on Public Works. *Water Pollution Control—1966: Hearings before the Subcommittee on Air and Water Pollution.* 89th Cong., 2d sess., April–May 1966.

Proceedings, Third Meeting of the Conference in the Matter of Pollution of Lake Erie and Its Tributaries, Cleveland, June 22, 1966. Washington, D.C.: FWPCA, 1966.

U.S. Congress. House Committee on Government Operations. *Water Pollution—Great Lakes (Part 1): Lake Ontario and Lake Erie: Hearings before the Subcommittee on Natural Resources and Power.* 89th Cong., 2d sess., July 22, 1966.

U.S. Congress. House Committee on Government Operations. *Water Pollution—Great Lakes (Part 3): Western Lake Erie, Detroit River, Lake St. Clair, and Tributaries: Hearings before the Subcommittee on Natural Resources and Power.* 89th Cong., 2d sess., September 9, 1966.

U.S. Congress. House Committee on Government Operations. *Water Pollution—Great Lakes (Part 4): Southwestern Lake Michigan: Milwaukee, Root, and Pike Rivers: Hearings before the Subcommittee on Natural Resources and Power.* 89th Cong., 2d sess., September 16, 1966.

Proceedings, Progress Meeting in the Matter of Pollution of the Interstate Waters of the Grand Calumet River, Little Calumet River, Calumet River, Wolf Lake, Lake Michigan, and Their Tributaries, Chicago, March 15, 1967. Washington, D.C.: FWPCA, 1967.

Proceedings, Third Session of the Conference in the Matter of Pollution of Lake Erie and Its Tributaries, Buffalo, March 22, 1967. Washington, D.C.: FWPCA, 1967.

U.S. Congress. Senate Committee on Public Works. *Water Pollution—1967 (Part 2): Hearings before the Subcommittee on Air and Water Pollution.* 90th Cong., 1st sess., August 9–10, 1967.

Proceedings, Progress Evaluation Meeting in the Matter of Pollution of the Interstate Waters of the Grand Calumet River, Little Calumet River, Calumet River, Wolf Lake, Lake Michigan and Their Tributaries, Chicago, September 11, 1967. Washington, D.C.: FWPCA, 1967.

Proceedings of the Conference in the Matter of Pollution of Lake Michigan and Its Tributary Basin, Chicago, January–March 1968. Washington, D.C.: FWPCA, 1968.

U.S. Congress. Senate Committee on Public Works. *Water Pollution—1968 (Part 1): Hearings before the Subcommittee on Air and Water Pollution.* 90th Cong., 2d sess., March 27, 1968.

Proceedings, Progress Evaluation Meeting of the Conference in the Matter of Pollution of Lake Erie and Its Tributaries, Cleveland, June 4, 1968. Washington, D.C.: FWPCA, 1968.

Proceedings, Technical Session of the Conference in the Matter of Pollution of Lake Erie and Its Tributaries, Cleveland, August 26, 1968. Washington, D.C.: FWPCA, 1969.

Proceedings, Fourth Session of the Conference in the Matter of Pollution of Lake Erie and Its Tributaries, Cleveland, October 4, 1968. Washington, D.C.: FWPCA, 1969.

Proceedings of the Second Session of the Conference in the Matter of Pollution of the Interstate Waters of the Grand Calumet River, Little Calumet River, Calumet River, Wolf Lake, Lake Michigan, and Their Tributaries, Chicago, December 11–12, 1968, and January 29, 1969. Washington, D.C.: FWPCA, 1969.

Proceedings, Second Session of the Conference in the Matter of Pollution of Lake Michigan and Its Tributary Basin, Chicago, February 25, 1969. Washington, D.C.: FWPCA, 1969.

Proceedings, Progress Evaluation Meeting of the Conference in the Matter of Pollution of Lake Erie and Its Tributaries, Cleveland, June 27, 1969. Washington, D.C.: FWPCA, 1970.

Proceedings, Reconvened Second Session of the Conference in the Matter of Pollution of the Interstate Waters of the Grand Calumet River, Little Calumet River, Calumet River, Wolf Lake, Lake Michigan, and Their Tributaries, Chicago, August 26, 1969. Washington, D.C.: FWPCA, 1970.

U.S. Congress. House Committee on Government Operations. *Phosphates in Detergents and the Eutrophication of America's Waters: Hearings before the Subcommittee on Conservation and Natural Resources.* 91st Cong., 1st sess., December 15, 16, 1969.

International Joint Commission. *Pollution of Lake Erie, Lake Ontario, and the International Section of the St. Lawrence River.* 3 vols. Washington, D.C.: IJC, Canada and the United States, 1970.

Proceedings, Third Session of the Conference in the Matter of Pollution of Lake Michigan and Its Tributary Basin, Milwaukee, March 31 and April 1, 1970. Washington, D.C.: FWQA, 1970.

U.S. Congress. Senate Committee on Public Works. *Water Pollution 1970 (Part 1): Hearings before the Subcommittee on Air and Water Pollution.* 91st Cong., 2d sess., April 20–21, 27, 1970.

U.S. Congress. Senate Committee on Public Works. *Water Pollution 1970 (Part 2): Hearings before the Subcommittee on Air and Water Pollution.* 91st Cong., 2d sess., April–May 1970.

Proceedings, Executive Session of the Third Session of the Conference in the Matter of Pollution of Lake Michigan and Its Tributary Basin, Chicago, May 7, 1970. Washington, D.C.: FWQA, 1970.

U.S. Congress. Senate Committee on Commerce. *Effects of Mercury on Man and the Environment (Part 1): Hearings before the Subcommittee on Energy, Natural Resources, and the Environment.* 91st Cong., 2d sess., May 8, 1970.

Proceedings, Fifth Session of the Conference in the Matter of Pollution of Lake Erie and Its Tributaries, Detroit, June 3–4, 1970. Washington, D.C.: FWQA, 1971.

U.S. Congress. Senate Committee on Public Works. *Water Pollution 1970 (Part 5): Hearings before the Subcommittee on Air and Water Pollution.* 91st Cong., 2d sess., June 9–10, 1970.

U.S. Congress. Senate Committee on Public Works. *Water Pollution Control Programs: Hearings before the Subcommittee on Air and Water Pollution.* 92d Cong., 1st sess., February 4, 8–9, 1971.

U.S. Congress. House Committee on Commerce. *Refuse Act Permit Program: Hearings before the Subcommittee on the Environment.* 92d Cong., 1st sess., February 18, 19, 1971.

U.S. Congress. Senate Committee on Public Works. *Water Pollution Control Legislation (Part 1): Hearings before the Subcommittee on Air and Water Pollution.* 92d Cong., 1st sess., March 1971.

U.S. Congress. Senate Committee on Public Works. *Water Pollution Control Legislation (Part 2): Hearings before the Subcommittee on Air and Water Pollution.* 92d Cong., 1st sess., March 1971.

Proceedings, Third Session (Reconvened) of the Conference in the Matter of Pollution of Lake Michigan and Its Tributary Basin, Chicago, March 23–25, 1971. Washington, D.C.: EPA, 1971.

U.S. Congress. House Committee on Public Works. *Water Pollution Control Legislation—1971 (Oversight of Existing Program).* 92d Cong., 1st sess., May–July 1971.

U.S. Congress. House Committee on Government Operations. *Mercury Pollution and Enforcement of the Refuse Act of 1899 (Part 1): Hearing before the Conservation and Natural Resources Subcommittee.* 92d Cong., 1st sess., July 1, 1971.

U.S. Congress. House Committee on Government Operations. *Phosphates and Phosphate Substitutes in Detergents (Parts 1 and 2): Hearings before the Subcommittee on Conservation and Natural Resources.* 92d Cong., 1st sess., October 20, 27–29, 1971

U.S. Congress. Senate Committee on Commerce. *The Toxic Substances Control Act of 1971 and Amendment: Hearings before the Subcommittee on the Environment.* 92d Cong., 1st sess., October 1, 15, 29, and November 5, 1971.

U.S. Congress. House Committee on Public Works. *Water Pollution Control Legislation—1971.* 92d Cong., 1st sess., December 7–10, 1971.

U.S. Environmental Protection Agency. *The First Two Years: A Review of EPA's Enforcement Program, February 1973.* Washington, D.C.: EPA, 1973.

U.S. Congress. House Committee on Public Works. *Implementation of the Federal Water Pollution Control Act: Hearings before the Subcommittee on Investigations and Review.* 93d Cong., 2d sess., February–June 1974.

U.S. Congress. House Committee on Public Works and Transportation. *Implementation of the Federal Water Pollution Control Act: Hearings before the Subcommittee on Investigations and Review.* 94th Cong., 2d sess., February 1976.

U.S. Congress. Senate Committee on Environment and Public Works. *Federal Water*

Pollution Control Act Amendments: Hearings before the Subcommittee on Environmental Pollution. 95th Cong., 1st sess., June 1977.

U.S. Congress. House Committee on Public Works and Transportation. *Implementation of the Federal Water Pollution Control Act: Hearings before the Subcommittee on Investigations and Review.* 95th Cong., 2d sess., June 1978.

Dissertations

Ayers, Michael Orin. "An Evaluation of Water Pollution Control Arrangements in and by the State of Illinois." Ph.D. diss., University of Oklahoma, 1974.

Dreisziger, Nandor Alexander Fred. "The International Joint Commission of the United States and Canada, 1895–1920: A Study in Canadian-American Relations." Ph.D. diss., University of Toronto, 1974.

Flannery, James Joseph. "Water Pollution Control: Development of State and National Policy." Ph.D. diss., University of Wisconsin, Madison, 1956.

Gargan, John Joseph. "The Politics of Water Pollution in New York State—The Development and Adoption of the 1965 Pure Waters Program." Ph.D. diss., Syracuse University, 1968.

Heidemann, Mary Ann. "Regional Ecology and Regulatory Federalism: Wisconsin's Quandary over Toxic Contamination of Green Bay." Ph.D. diss., University of Wisconsin—Madison, 1989.

Huffman, Thomas R. "Protectors of the Land and Water: The Political Culture of Conservation and the Rise of Environmentalism in Wisconsin, 1958–1970." Ph.D. diss., University of Wisconsin—Madison, 1989.

Hurley, Andrew J. "Environmental and Social Change in Gary, Indiana, 1945–1980." Ph.D. diss., Northwestern University, 1988.

Secondary Sources

Advisory Commission on Intergovernmental Relations. *Regulatory Federalism: Policy, Process, Impact, and Reform.* Washington, D.C.: GPO, 1984.

Allan, Leslie, et al. *Council on Economic Priorities. Paper Profits: Pollution in the Pulp and Paper Industry.* Cambridge, Mass.: MIT Press, 1972.

Ashworth, William. *The Late, Great Lakes.* New York: Alfred Knopf, 1986.

Balogh, Brian. *Chain Reaction: Expert Debate and Public Participation in American Commercial Nuclear Power, 1945–1975.* Cambridge, England: Cambridge University Press, 1991.

Bartlett, Robert V. *The Reserve Mining Controversy: Science, Technology, and Environmental Quality.* Bloomington: Indiana University Press, 1980.

Bastow, Thomas F. *"This Vast Pollution . . .": United States of America vs. Reserve Mining Company.* Washington, D.C.: Green Fields Books, 1986.

Beyle, Thad L., Thomas E. Peddicord, and Francis H. Parker. *Integration and Coordination of State Environmental Programs.* Lexington, Ky.: Council of State Governments, 1975.

Brown, Theodore L. *Energy and the Environment.* Columbus, Ohio: Charles E. Merrill Publishing, 1971.

Cain, Louis P. *Sanitation Strategy for a Lakefront Metropolis: The Case of Chicago.* DeKalb: Northern Illinois University Press, 1978.

Cannon, James S., and Frederick S. Armentrout. *Environmental Steel Update: Pollution in the Iron and Steel Industry.* New York: Council on Economic Priorities, 1977.

Cannon, James S., and Jean M. Halloran, for the Council on Economic Priorities. *Environmental Steel: Pollution in the Iron and Steel Industry.* New York: Praeger, 1974.

Carrol, John E. *Environmental Diplomacy: An Examination and a Prospective of Canadian-U.S. Transboundary Environmental Relations.* Ann Arbor: University of Michigan Press, 1983.

Childs, William R. *Trucking and the Public Interest: The Emergence of Federal Regulation, 1914–1940.* Knoxville: University of Tennessee Press, 1985.

Colborn, Theodora, et al. *Great Lakes, Great Legacy?* Washington, D.C.: Conservation Foundation, 1990.

Cowdrey, Albert E. "Pioneering Environmental Law: The Army Corps of Engineers and the Refuse Act." *Pacific Historical Review* 44 (August 1975): 331–49.

Crandall, Robert W. "The Politics of the Environment, 1970–1987: Learning the Lessons." *Wilson Quarterly* 11, no. 4 (autumn 1987): 69–80.

Derthick, Martha. "Crossing Thresholds: Federalism in the 1960s." *Journal of Policy History* 8, no. 1 (1996): 64–80.

D'Itri, Patricia A., and Frank M. D'Itri. *Mercury Contamination: A Human Tragedy.* New York: John Wiley and Sons, 1977.

Downing, Paul B., and James N. Kimball. "Enforcing Pollution Control Laws in the U.S." *Policy Studies Journal* 11, no. 1 (September 1982): 55–65.

Duffy, John. *The Sanitarians: A History of American Public Health.* Urbana: University of Illinois Press, 1990.

Dunlap, Riley E. "Trends in Public Opinion toward Environmental Issues, 1965–1990." In *American Environmentalism: The U.S. Environmental Movement, 1970–1990,* ed. Riley E. Dunlap and Angela G. Mertig, 89–117. New York: Taylor and Francis, 1992.

Dunlap, Thomas R. *DDT: Scientists, Citizens, and Public Policy.* Princeton: Princeton University Press, 1981.

Ebbin, Steve, and Ronald Kasper. *Citizen Groups and the Nuclear Power Industry: Uses of Scientific and Technological Information.* Cambridge, Mass.: MIT Press, 1974.

Egerton, Frank N. "Missed Opportunities: U.S. Fishery Biologists and Productivity of Fish in Green Bay, Saginaw Bay, and Western Lake Erie." *Environmental Review* 13, no. 2 (summer 1989): 33–63.

———. "Pollution and Aquatic Life in Lake Erie: Early Scientific Studies." *Environmental Review* 11, no. 3 (fall 1987): 189–205.

Eisner, Marc Allen. "Discovering Patterns in Regulatory History: Continuity, Change, and Regulatory Regimes." *Journal of Policy History* 6, no. 2 (1994): 175–79.

Elazar, Daniel J. *American Federalism: A View from the States.* 3d ed. New York: Harper and Row, 1984.

Fox, Stephen. *John Muir and His Legacy: The American Conservation Movement.* Boston: Little, Brown, 1981.

Galambos, Louis. "The Emerging Organizational Synthesis in Modern American History." *Business History Review* 44 (autumn 1970): 280–90.

———. "Technology, Political Economy, and Professionalization: Central Themes of the Organizational Synthesis." *Business History Review* 57 (winter 1983): 471–93.

Galishoff, Stuart. "Triumph and Failure: The American Response to the Urban Water Supply Problem, 1860–1923." In *Pollution and Reform in American Cities,* ed. Martin V. Melosi, pp. 35–58. Austin: University of Texas Press, 1980.

Gottlieb, Robert. *Forcing the Spring: The Transformation of the American Environmental Movement.* Washington, D.C.: Island Press, 1993.

Graham, Hugh Davis. "The Stunted Career of Policy History: A Critique and an Agenda." *Public Historian* 15, no. 2 (spring 1993): 22–26.

Graham, Otis L., Jr. *Toward a Planned Society: From Roosevelt to Nixon.* New York: Oxford University Press, 1976.

Griffith, Robert. "Dwight D. Eisenhower and the Corporate Commonwealth." *American Historical Review* 87 (February 1982): 87–122.

Haskell, Elizabeth H., and Victoria S. Price. *State Environmental Management: Case Studies of Nine States.* New York: Praeger, 1973.

Haskell, Thomas L. "Introduction." In *The Authority of Experts: Studies in History and Theory,* ed. Haskell, pp. ix–xxxix. Bloomington: Indiana University Press, 1984.

Hawley, Ellis W. "The Discovery and Study of a 'Corporate Liberalism'." *Business History Review* 52 (autumn 1978): 309–20.

———. "Social Policy and the Liberal State in Twentieth-Century America." In *Federal Social Policy: The Historical Dimension,* ed. Donald T. Critchlow and Hawley, pp. 117–39. State College: Pennsylvania State University Press, 1988.

Hays, Samuel P. *Beauty, Health, and Permanence: Environmental Politics in the United States, 1955–1985.* Cambridge, England: Cambridge University Press, 1987.

———. *Conservation and the Gospel of Efficiency: The Progressive Conservation Movement, 1890–1920.* Cambridge, Mass.: Harvard University Press, 1959.

———. "From Conservation to Environment: Environmental Politics in the United States since World War Two." *Environmental Review* 6, no. 2 (fall 1982): 14–41.

———. "Political Choice in Regulatory Administration." In *Regulation in Perspective: Historical Essays,* ed. Thomas McCraw, pp. 124–54. Cambridge, Mass.: Harvard University Press, 1981.

Heclo, Hugh. "The Sixties' False Dawn: Awakenings, Movements, and Postmodern Policymaking." *Journal of Policy History* 8, no. 1 (1996): 34–63.

Hoberg, George. *Pluralism by Design: Environmental Policy and the American Regulatory State.* Westport, Conn.: Praeger, 1992.

Hogan, Michael J. "Corporatism." In "A Round Table: Explaining the History of American Foreign Relations." *Journal of American History* 77, no. 1 (June 1990): 153–60.

Holmes, Beatrice Hort. *History of Federal Water Resources Programs and Policies, 1961–1970.* Washington, D.C.: U.S. Department of Agriculture, 1979.

Hunciker, Kurt M. "The Clean Water Act and Related Developments in the Federal Water Pollution Control Program during 1977 (Part 2): Modifications of the Municipal Program." *Harvard Environmental Law Review* 2 (1977): 127–75.

Kerlin, Gregg, and Daniel Rabovsky. *Cracking Down: Oil Refining and Pollution Control.* New York: Council on Economic Priorities, 1975.

Komanoff, Charles, Holly Miller, and Sandy Noyes, *The Price of Power: Electric Utilities and the Environment.* New York: Council on Economic Priorities, 1972.

Lacey, Michael J., ed. *Government and Environmental Politics: Essays on Historical Developments since World War Two.* Washington, D.C.: Woodrow Wilson Center Press, 1989.

Larson, Magali Sarfatti. "The Production of Expertise and the Constitution of Expert Power." In *The Authority of Experts*, ed. Haskell, pp. 28–83. Bloomington: Indiana University Press, 1984.

Leuchtenberg, William E. "The Pertinence of Political History: Reflections on the Significance of the State in America." *Journal of American History*, 73 (December 1986): 585–600.

Lewis, Richard S. *The Nuclear-Power Rebellion: Citizens vs. the Atomic Industrial Establishment*. New York: Viking Press, 1972.

Lieber, Harvey. *Federalism and Clean Waters: The 1972 Water Pollution Control Act*. Lexington, Mass.: Lexington Books, D. C. Heath and Company, 1975.

Lowi, Theodore J. *The End of Liberalism: The Second Republic of the United States*. 2d ed. New York: W. W. Norton, 1979.

McCraw, Thomas K. "Regulation in America: A Review Article." *Business History Review* 49 (summer 1975): 159–83.

McConnel, Grant. *Private Power and American Democracy*. New York: Alfred A. Knopf, 1967.

McGucken, William. *Biodegradable: Detergents and the Environment*. College Station: Texas A&M University Press, 1991.

———. "The Canadian Federal Government, Cultural Eutrophication, and the Regulation of Detergent Phosphates, 1970." *Environmental Review* 13, no. 3–4 (fall–winter 1989): 155–66.

Marcus, Alfred A. *The Adversary Economy: Business Responses to Changing Government Requirements*. Westport, Conn.: Quorum Books, 1984.

———. *Promise and Performance: Choosing and Implementing an Environmental Policy*. Westport, Conn.: Greenwood Press, 1980.

Martin, Roscoe C. *Water for New York: A Study in State Administration of Water Resources*. Syracuse: Syracuse University Press, 1960.

Melosi, Martin. *Coping with Abundance: Energy and Environment in Industrial America*. Philadelphia: Temple University Press, 1985.

———. "Lyndon Johnson and Environmental Policy." In *The Johnson Years*: vol. 2, *Vietnam, the Environment, and Science*, ed. Robert A. Divine, pp. 113–49. Lawrence: University Press of Kansas, 1987.

Micklin, Philip P. "Water Quality: A Question of Standards." In *Congress and the Environment*, ed. Richard A. Cooley and Geoffrey Wandesforde-Smith, pp. 130–47. Seattle: University of Washington Press, 1970.

Mitchell, Robert Cameron. "From Conservation to Environmental Movement: The Development of the Modern Environmental Lobbies." In *Government and Environmental Politics: Essays on Historical Developments since World War Two*, ed. Michael J. Lacey, pp. 81–113. Washington, D.C.: Woodrow Wilson Center Press, 1989.

Munton, Don. "Dependence and Interdependence in Transboundary Environmental Relations." *International Journal* 36, no. 1 (winter 1980–1981): 139–84.

Murphy, Earl Finbar. *Water Purity: A Study in Legal Control of Natural Resources*. Madison: University of Wisconsin Press, 1961.

O'Brien, Jim. "Environmentalism as a Mass Movement: Historical Notes." *Radical America* 17, no. 2–3 (March–June 1983): 7–27.

Petulla, Joseph M. *Environmental Protection in the United States: Industry, Agencies, Environmentalists*. San Francisco: San Francisco Study Center, 1987.

Quarles, John. *Cleaning Up America: An Insider's View of the Environmental Protection*

Agency. Boston: Houghton Mifflin, 1976.

Rabin, Robert L. "Federal Regulation in Historical Perspective." *Stanford Law Review* 38 (1986): 1189–1326.

Rolph, Elizabeth S. *Nuclear Power and the Public Safety.* Lexington, Mass.: Lexington Books, 1979.

Rosenbaum, Walter. *Environmental Politics and Policy.* Washington, D.C.: Congressional Quarterly Press, 1985 (1st ed.); 1995 (3d ed.).

Rukeyser, William Simon. "Fact and Foam in the Row over Phosphates." *Fortune* (January 1972): 72–73.

Sabato, Larry. *Goodbye to Good-Time Charlie: The American Governor Transformed, 1950–1975.* Lexington, Mass.: Lexington Books, D. C. Heath and Company, 1978.

Sale, Kirkpatrick. *The Green Revolution: The American Environmental Movement, 1962–1992.* New York: Hill and Wang, 1993.

Scarpino, Philip V. *Great River: An Environmental History of the Upper Mississippi, 1890–1950.* Columbia: University of Missouri Press, 1985.

Schmandt, Jurgen, Ernest T. Smerdon, and Judith Clarkson. *State Water Policies: A Study of Six States.* New York: Praeger, 1988.

Seely, Bruce E. *Building the American Highway System: Engineers as Policy Makers.* Philadelphia: Temple University Press, 1987.

Sirgo, Henry B. "Water Policy Decision-Making and Implementation in the Johnson Administration." *Journal of Political Science* 12, no. 1–2 (1985): 53–63.

Tarr, Joel A. "Historical Perspectives on Hazardous Wastes in the United States." *Waste Management and Research* 3 (1985): 95–102.

———. "Industrial Wastes and Public Health: Some Historical Notes (Part 1, 1876–1932)." *American Journal of Public Health* 75, no. 9 (September 1985): 1059–67.

Tarr, Joel A., James McCurley, and Terry F. Yosie. "The Development and Impact of Urban Wastewater Technology: Changing Concepts of Water Quality Control, 1850–1930." In *Pollution and Reform in American Cities,* ed. Martin V. Melosi, pp. 59–82. Austin: University of Texas Press, 1980.

Tarr, Joel A., Terry Yosie, and James McCurley III. "Disputes over Water Quality Policy: Professional Cultures in Conflict, 1900–1917." *American Journal of Public Health* 70, no. 4 (April 1980): 427–35.

Tarr, Joel A., et al. *Retrospective Assessment of Wastewater Technology in the United States, 1800–1972: A Report to the National Science Foundation.* Carnegie-Mellon University, 1977.

Vogel, David. *Fluctuating Fortunes: The Political Power of Business in America.* New York: Basic Books, 1989.

———. "The 'New' Social Regulation in Historical and Comparative Perspective." In *Regulation in Perspective,* ed. Thomas McCraw, pp. 155–86. Cambridge, Mass.: Harvard University Press, 1981.

———. "The Public-Interest Movement and the American Reform Tradition." *Political Science Quarterly* 95, no. 4 (winter 1980–1981): 607–27.

Walker, J. Samuel. *Containing the Atom: Nuclear Regulation in a Changing Environment, 1963–1971.* Berkeley and Los Angeles: University of California Press, 1992.

———. "Nuclear Power and the Environment: The Atomic Energy Commission and Thermal Pollution, 1965–1971." *Technology and Culture* 30, no. 4 (October

1989): 964–92.

Wenner, Lettie M. *The Environmental Decade in Court.* Bloomington: Indiana University Press, 1982.

Whitaker, John C. *Striking a Balance: Environment and Natural Resources Policy in the Nixon-Ford Years.* Washington, D.C.: American Enterprise Institute for Public Policy Research, 1976.

Wilson, James Q. *Bureaucracy: What Government Agencies Do and Why They Do It.* New York: Basic Books, 1989.

———. "The Politics of Regulation." In *The Politics of Regulation,* ed. Wilson, pp. 357–94. New York: Basic Books, 1980.

Zwick, David R., and Marcy Benstock. *Water Wasteland: Ralph Nader's Study Group Report on Water Pollution.* New York: Grossman, 1971.

Index